GNU PSPP Reference Manual

A catalogue record for this book is available from the Hong Kong Public Libraries.

Published in Hong Kong by Samurai Media Limited.

Email: info@samuraimedia.org

ISBN 978-988-8381-44-9

Background Cover Image by https://www.flickr.com/people/webtreatsetc/

Table of Contents

1 Introduction 2

2 Your rights and obligations 3

3 Invoking pspp 4
   3.1 Main Options 4
   3.2 PDF, PostScript, and SVG Output Options 7
   3.3 Plain Text Output Options 8
   3.4 HTML Output Options 9
   3.5 OpenDocument Output Options 10
   3.6 Comma-Separated Value Output Options 10

4 Invoking psppire 12
   4.1 The graphic user interface 12

5 Using PSPP 13
   5.1 Preparation of Data Files 13
      5.1.1 Defining Variables 14
      5.1.2 Listing the data 15
      5.1.3 Reading data from a text file 15
      5.1.4 Reading data from a pre-prepared PSPP file 15
      5.1.5 Saving data to a PSPP file 16
      5.1.6 Reading data from other sources 16
      5.1.7 Exiting PSPP 16
   5.2 Data Screening and Transformation 16
      5.2.1 Identifying incorrect data 16
      5.2.2 Dealing with suspicious data 18
      5.2.3 Inverting negatively coded variables 19
      5.2.4 Testing data consistency 19
      5.2.5 Testing for normality 20
   5.3 Hypothesis Testing 23
      5.3.1 Testing for differences of means 23
      5.3.2 Linear Regression 24

6 The PSPP language 28
   6.1 Tokens 28
   6.2 Forming commands of tokens 29
   6.3 Syntax Variants 30
   6.4 Types of Commands 30
   6.5 Order of Commands 31
   6.6 Handling missing observations 32

6.7   Datasets 32
   6.7.1   Attributes of Variables 32
   6.7.2   Variables Automatically Defined by PSPP 34
   6.7.3   Lists of variable names 34
   6.7.4   Input and Output Formats 34
      6.7.4.1   Basic Numeric Formats............................ 35
      6.7.4.2   Custom Currency Formats......................... 37
      6.7.4.3   Legacy Numeric Formats 38
      6.7.4.4   Binary and Hexadecimal Numeric Formats.......... 39
      6.7.4.5   Time and Date Formats 40
      6.7.4.6   Date Component Formats 43
      6.7.4.7   String Formats 43
   6.7.5   Scratch Variables.. 43
6.8   Files Used by PSPP... 43
6.9   File Handles 44
6.10   Backus-Naur Form 45

7   Mathematical Expressions 46
7.1   Boolean Values 46
7.2   Missing Values in Expressions 46
7.3   Grouping Operators 46
7.4   Arithmetic Operators 46
7.5   Logical Operators.. 47
7.6   Relational Operators 47
7.7   Functions 48
   7.7.1   Mathematical Functions................................... 48
   7.7.2   Miscellaneous Mathematical Functions.................... 48
   7.7.3   Trigonometric Functions................................... 49
   7.7.4   Missing-Value Functions................................... 49
   7.7.5   Set-Membership Functions 50
   7.7.6   Statistical Functions....................................... 50
   7.7.7   String Functions 51
   7.7.8   Time & Date Functions 53
      7.7.8.1   How times & dates are defined and represented 53
      7.7.8.2   Functions that Produce Times 53
      7.7.8.3   Functions that Examine Times 53
      7.7.8.4   Functions that Produce Dates 54
      7.7.8.5   Functions that Examine Dates..................... 55
      7.7.8.6   Time and Date Arithmetic 56
   7.7.9   Miscellaneous Functions................................... 57
   7.7.10   Statistical Distribution Functions....................... 57
      7.7.10.1   Continuous Distributions 58
      7.7.10.2   Discrete Distributions 62
7.8   Operator Precedence 63

8   Data Input and Output 64

8.1   BEGIN DATA 64
8.2   CLOSE FILE HANDLE 64
8.3   DATAFILE ATTRIBUTE 64
8.4   DATASET commands 65
8.5   DATA LIST 66
    8.5.1   DATA LIST FIXED 66
        Examples 68
    8.5.2   DATA LIST FREE 69
    8.5.3   DATA LIST LIST 70
8.6   END CASE 70
8.7   END FILE 70
8.8   FILE HANDLE 70
8.9   INPUT PROGRAM 73
8.10   LIST 76
8.11   NEW FILE 76
8.12   PRINT 76
8.13   PRINT EJECT 77
8.14   PRINT SPACE 78
8.15   REREAD 78
8.16   REPEATING DATA 78
8.17   WRITE 80

9   System and Portable File I/O 81

9.1   APPLY DICTIONARY 81
9.2   EXPORT 82
9.3   GET 82
9.4   GET DATA 83
    9.4.1   Spreadsheet Files 84
    9.4.2   Postgres Database Queries 84
    9.4.3   Textual Data Files 85
        9.4.3.1   Reading Delimited Data 86
        9.4.3.2   Reading Fixed Columnar Data 88
9.5   IMPORT 89
9.6   SAVE 89
9.7   SAVE TRANSLATE 91
    9.7.1   Writing Comma- and Tab-Separated Data Files 92
9.8   SYSFILE INFO 93
9.9   XEXPORT 93
9.10   XSAVE 94

10   Combining Data Files 95

10.1   Common Syntax 95
10.2   ADD FILES 97
10.3   MATCH FILES 98
10.4   UPDATE 99

11 Manipulating variables **100**

11.1 ADD VALUE LABELS 100
11.2 DELETE VARIABLES 100
11.3 DISPLAY 100
11.4 FORMATS 101
11.5 LEAVE 101
11.6 MISSING VALUES 102
11.7 MODIFY VARS 103
11.8 MRSETS 103
11.9 NUMERIC 105
11.10 PRINT FORMATS 105
11.11 RENAME VARIABLES 105
11.12 VALUE LABELS 105
11.13 STRING 106
11.14 VARIABLE ATTRIBUTE 106
11.15 VARIABLE LABELS 107
11.16 VARIABLE ALIGNMENT 107
11.17 VARIABLE WIDTH 108
11.18 VARIABLE LEVEL 108
11.19 VARIABLE ROLE 108
11.20 VECTOR 109
11.21 WRITE FORMATS 109

12 Data transformations **110**

12.1 AGGREGATE 110
12.2 AUTORECODE 113
12.3 COMPUTE 113
12.4 COUNT 114
12.5 FLIP 115
12.6 IF 116
12.7 RECODE 116
12.8 SORT CASES 119

13 Selecting data for analysis **120**

13.1 FILTER 120
13.2 N OF CASES 120
13.3 SAMPLE 121
13.4 SELECT IF 121
13.5 SPLIT FILE 121
13.6 TEMPORARY 122
13.7 WEIGHT 123

14 Conditional and Looping Constructs **124**

14.1 BREAK 124
14.2 DO IF 124
14.3 DO REPEAT 124
14.4 LOOP 125

15   Statistics **127**

15.1   DESCRIPTIVES 127
15.2   FREQUENCIES 128
15.3   EXAMINE 130
15.4   GRAPH 132
15.5   CORRELATIONS 132
15.6   CROSSTABS 133
15.7   FACTOR 136
15.8   LOGISTIC REGRESSION 138
15.9   MEANS 139
15.10   NPAR TESTS 141
   15.10.1   Binomial test 141
   15.10.2   Chisquare Test 142
   15.10.3   Cochran Q Test 142
   15.10.4   Friedman Test 142
   15.10.5   Kendall's W Test 143
   15.10.6   Kolmogorov-Smirnov Test 143
   15.10.7   Kruskal-Wallis Test 143
   15.10.8   Mann-Whitney U Test 144
   15.10.9   McNemar Test 144
   15.10.10   Median Test 144
   15.10.11   Runs Test 144
   15.10.12   Sign Test 145
   15.10.13   Wilcoxon Matched Pairs Signed Ranks Test 145
15.11   T-TEST 145
   15.11.1   One Sample Mode 146
   15.11.2   Independent Samples Mode 146
   15.11.3   Paired Samples Mode 146
15.12   ONEWAY 147
15.13   QUICK CLUSTER 148
15.14   RANK 148
15.15   REGRESSION 149
   15.15.1   Syntax 150
   15.15.2   Examples 150
15.16   RELIABILITY 151
15.17   ROC 151

16   Utilities **153**

16.1   ADD DOCUMENT 153
16.2   CACHE 153
16.3   CD 153
16.4   COMMENT 153
16.5   DOCUMENT 153
16.6   DISPLAY DOCUMENTS 154
16.7   DISPLAY FILE LABEL 154
16.8   DROP DOCUMENTS 154
16.9   ECHO 154
16.10   ERASE 154

16.11   EXECUTE... 154

16.12   FILE LABEL.. 155

16.13   FINISH 155

16.14   HOST.. 155

16.15   INCLUDE.. 155

16.16   INSERT.. 155

16.17   OUTPUT 156

16.18   PERMISSIONS....................................... 157

16.19   PRESERVE and RESTORE............................. 157

16.20   SET... 157

16.21   SHOW 164

16.22   SUBTITLE... 165

16.23   TITLE 165

17   Invoking `pspp-convert` **166**

18   Invoking `pspp-dump-sav` **168**

19   Not Implemented **169**

20   Bugs **174**

20.1 When to report bugs....................................... 174

20.2 How to report bugs 174

21   Function Index **176**

22   Command Index **179**

23   Concept Index **181**

Appendix A   GNU Free Documentation License
    .... **186**

1 Introduction

PSPP is a tool for statistical analysis of sampled data. It reads the data, analyzes the data according to commands provided, and writes the results to a listing file, to the standard output or to a window of the graphical display.

The language accepted by PSPP is similar to those accepted by SPSS statistical products. The details of PSPP's language are given later in this manual.

PSPP produces tables and charts as output, which it can produce in several formats; currently, ASCII, PostScript, PDF, HTML, and DocBook are supported.

The current version of PSPP, 0.8.5, is incomplete in terms of its statistical procedure support. PSPP is a work in progress. The authors hope to fully support all features in the products that PSPP replaces, eventually. The authors welcome questions, comments, donations, and code submissions. See Chapter 20 [Submitting Bug Reports], page 174, for instructions on contacting the authors.

2 Your rights and obligations

PSPP is not in the public domain. It is copyrighted and there are restrictions on its distribution, but these restrictions are designed to permit everything that a good cooperating citizen would want to do. What is not allowed is to try to prevent others from further sharing any version of this program that they might get from you.

Specifically, we want to make sure that you have the right to give away copies of PSPP, that you receive source code or else can get it if you want it, that you can change these programs or use pieces of them in new free programs, and that you know you can do these things.

To make sure that everyone has such rights, we have to forbid you to deprive anyone else of these rights. For example, if you distribute copies of PSPP, you must give the recipients all the rights that you have. You must make sure that they, too, receive or can get the source code. And you must tell them their rights.

Also, for our own protection, we must make certain that everyone finds out that there is no warranty for PSPP. If these programs are modified by someone else and passed on, we want their recipients to know that what they have is not what we distributed, so that any problems introduced by others will not reflect on our reputation.

Finally, any free program is threatened constantly by software patents. We wish to avoid the danger that redistributors of a free program will individually obtain patent licenses, in effect making the program proprietary. To prevent this, we have made it clear that any patent must be licensed for everyone's free use or not licensed at all.

The precise conditions of the license for PSPP are found in the GNU General Public License. You should have received a copy of the GNU General Public License along with this program; if not, write to the Free Software Foundation, Inc., 51 Franklin Street, Fifth Floor, Boston, MA 02110-1301 USA. This manual specifically is covered by the GNU Free Documentation License (see Appendix A [GNU Free Documentation License], page 186).

3 Invoking `pspp`

PSPP has two separate user interfaces. This chapter describes `pspp`, PSPP's command-line driven text-based user interface. The following chapter briefly describes PSPPIRE, the graphical user interface to PSPP.

The sections below describe the `pspp` program's command-line interface.

3.1 Main Options

Here is a summary of all the options, grouped by type, followed by explanations in the same order.

In the table, arguments to long options also apply to any corresponding short options.

Non-option arguments
> `syntax-file`

Output options
> `-o, --output=output-file`
> `-O option=value`
> `-O format=format`
> `-O device={terminal|listing}`
> `--no-output`
> `-e, --error-file=error-file`

Language options
> `-I, --include=dir`
> `-I-, --no-include`
> `-b, --batch`
> `-i, --interactive`
> `-r, --no-statrc`
> `-a, --algorithm={compatible|enhanced}`
> `-x, --syntax={compatible|enhanced}`
> `--syntax-encoding=encoding`

Informational options
> `-h, --help`
> `-V, --version`

Other options
> `-s, --safer`
> `--testing-mode`

syntax-file Read and execute the named syntax file. If no syntax files are specified, PSPP prompts for commands. If any syntax files are specified, PSPP by default exits after it runs them, but you may make it prompt for commands by specifying '-' as an additional syntax file.

`-o output-file`
> Write output to *output-file*. PSPP has several different output drivers that support output in various formats (use `--help` to list the available formats).

Specify this option more than once to produce multiple output files, presumably in different formats.

Use '-' as *output-file* to write output to standard output.

If no -o option is used, then PSPP writes text and CSV output to standard output and other kinds of output to whose name is based on the format, e.g. `pspp.pdf` for PDF output.

-O *option=value*

Sets an option for the output file configured by a preceding -o. Most options are specific to particular output formats. A few options that apply generically are listed below.

-O `format=`*format*

PSPP uses the extension of the file name given on -o to select an output format. Use this option to override this choice by specifying an alternate format, e.g. -o `pspp.out` -O `html` to write HTML to a file named `pspp.out`. Use `--help` to list the available formats.

-O `device={terminal|listing}`

Sets whether PSPP considers the output device configured by the preceding -o to be a terminal or a listing device. This affects what output will be sent to the device, as configured by the SET command's output routing subcommands (see Section 16.20 [SET], page 157). By default, output written to standard output is considered a terminal device and other output is considered a listing device.

--no-output

Disables output entirely, if neither -o nor -O is also used. If one of those options is used, --no-output has no effect.

-e *error-file*

--error-file=*error-file*

Configures a file to receive PSPP error, warning, and note messages in plain text format. Use '-' as *error-file* to write messages to standard output. The default error file is standard output in the absence of these options, but this is suppressed if an output device writes to standard output (or another terminal), to avoid printing every message twice. Use 'none' as *error-file* to explicitly suppress the default.

-I *dir*

--include=*dir*

Appends *dir* to the set of directories searched by the INCLUDE (see Section 16.15 [INCLUDE], page 155) and INSERT (see Section 16.16 [INSERT], page 155) commands.

-I-

--no-include

Clears all directories from the include path, including directories inserted in the include path by default. The default include path is . (the current directory), followed by .pspp in the user's home directory, followed by PSPP's system configuration directory (usually `/etc/pspp` or `/usr/local/etc/pspp`).

`-b`

`--batch`

`-i`
`--interactive`

> These options forces syntax files to be interpreted in batch mode or interactive mode, respectively, rather than the default "auto" mode. See Section 6.3 [Syntax Variants], page 30, for a description of the differences.

`-r`
`--no-statrc`

> Disables running `rc` at PSPP startup time.

`-a {enhanced|compatible}`
`--algorithm={enhanced|compatible}`

> With `enhanced`, the default, PSPP uses the best implemented algorithms for statistical procedures. With `compatible`, however, PSPP will in some cases use inferior algorithms to produce the same results as the proprietary program SPSS.

> Some commands have subcommands that override this setting on a per command basis.

`-x {enhanced|compatible}`
`--syntax={enhanced|compatible}`

> With `enhanced`, the default, PSPP accepts its own extensions beyond those compatible with the proprietary program SPSS. With `compatible`, PSPP rejects syntax that uses these extensions.

`--syntax-encoding=encoding`

> Specifies *encoding* as the encoding for syntax files named on the command line. The *encoding* also becomes the default encoding for other syntax files read during the PSPP session by the `INCLUDE` and `INSERT` commands. See Section 16.16 [INSERT], page 155, for the accepted forms of *encoding*.

`--help`   Prints a message describing PSPP command-line syntax and the available device formats, then exits.

`-V`
`--version`

> Prints a brief message listing PSPP's version, warranties you don't have, copying conditions and copyright, and e-mail address for bug reports, then exits.

`-s`
`--safer`   Disables certain unsafe operations. This includes the `ERASE` and `HOST` commands, as well as use of pipes as input and output files.

`--testing-mode`

> Invoke heuristics to assist with testing PSPP. For use by `make check` and similar scripts.

3.2 PDF, PostScript, and SVG Output Options

To produce output in PDF, PostScript, and SVG formats, specify `-o` *file* on the PSPP command line, optionally followed by any of the options shown in the table below to customize the output format.

PDF, PostScript, and SVG output is only available if your installation of PSPP was compiled with the Cairo library.

`-O format={pdf|ps|svg}`

> Specify the output format. This is only necessary if the file name given on `-o` does not end in `.pdf`, `.ps`, or `.svg`.

`-O paper-size=`*paper-size*

> Paper size, as a name (e.g. `a4`, `letter`) or measurements (e.g. `210x297`, `8.5x11in`).

> The default paper size is taken from the `PAPERSIZE` environment variable or the file indicated by the `PAPERCONF` environment variable, if either variable is set. If not, and your system supports the `LC_PAPER` locale category, then the default paper size is taken from the locale. Otherwise, if `/etc/papersize` exists, the default paper size is read from it. As a last resort, A4 paper is assumed.

`-O foreground-color=`*color*
`-O background-color=`*color*

> Sets *color* as the color to be used for the background or foreground. Color should be given in the format `#RRRRGGGGBBBB`, where *RRRR*, *GGGG* and *BBBB* are 4 character hexadecimal representations of the red, green and blue components respectively.

`-O orientation=`*orientation*

> Either `portrait` or `landscape`. Default: `portrait`.

`-O left-margin=`*dimension*
`-O right-margin=`*dimension*
`-O top-margin=`*dimension*
`-O bottom-margin=`*dimension*

> Sets the margins around the page. See below for the allowed forms of *dimension* Default: `0.5in`.

`-O prop-font=`*font-name*
`-O emph-font=`*font-name*
`-O fixed-font=`*font-name*

> Sets the font used for proportional, emphasized, or fixed-pitch text. Most systems support CSS-like font names such as "serif" and "monospace", but a wide range of system-specific font are likely to be supported as well.

> Default: proportional font `serif`, emphasis font `serif italic`, fixed-pitch font `monospace`.

`-O font-size=`*font-size*

> Sets the size of the default fonts, in thousandths of a point. Default: 10000 (10 point).

`-O line-gutter=dimension`
> Sets the width of white space on either side of lines that border text or graphics objects. Default: `1pt`.

`-O line-spacing=dimension`
> Sets the spacing between the lines in a double line in a table. Default: `1pt`.

`-O line-width=dimension`
> Sets the width of the lines used in tables. Default: `0.5pt`.

Each *dimension* value above may be specified in various units based on its suffix: '`mm`' for millimeters, '`in`' for inches, or '`pt`' for points. Lacking a suffix, numbers below 50 are assumed to be in inches and those about 50 are assumed to be in millimeters.

3.3 Plain Text Output Options

PSPP can produce plain text output, drawing boxes using ASCII or Unicode line drawing characters. To produce plain text output, specify `-o file` on the PSPP command line, optionally followed by options from the table below to customize the output format.

Plain text output is encoded in UTF-8.

`-O format=txt`
> Specify the output format. This is only necessary if the file name given on `-o` does not end in `.txt` or `.list`.

`-O charts={template.png|none}`
> Name for chart files included in output. The value should be a file name that includes a single '`#`' and ends in `png`. When a chart is output, the '`#`' is replaced by the chart number. The default is the file name specified on `-o` with the extension stripped off and replaced by `-#.png`.
>
> Specify `none` to disable chart output. Charts are always disabled if your installation of PSPP was compiled without the Cairo library.

`-O foreground-color=color`
`-O background-color=color`
> Sets *color* as the color to be used for the background or foreground to be used for charts. Color should be given in the format *#RRRRGGGGBBBB*, where *RRRR*, *GGGG* and *BBBB* are 4 character hexadecimal representations of the red, green and blue components respectively. If charts are disabled, this option has no effect.

`-O paginate=boolean`
> If set, PSPP writes an ASCII formfeed the end of every page. Default: `off`.

`-O headers=boolean`
> If enabled, PSPP prints two lines of header information giving title and subtitle, page number, date and time, and PSPP version are printed at the top of every page. These two lines are in addition to any top margin requested. Default: `off`.

`-O length=line-count`

> Physical length of a page. Headers and margins are subtracted from this value. You may specify the number of lines as a number, or for screen output you may specify `auto` to track the height of the terminal as it changes. Default: `66`.

`-O width=character-count`

> Width of a page, in characters. Margins are subtracted from this value. For screen output you may specify `auto` in place of a number to track the width of the terminal as it changes. Default: `79`.

`-O top-margin=top-margin-lines`

> Length of the top margin, in lines. PSPP subtracts this value from the page length. Default: `0`.

`-O bottom-margin=bottom-margin-lines`

> Length of the bottom margin, in lines. PSPP subtracts this value from the page length. Default: `0`.

`-O box={ascii|unicode}`

> Sets the characters used for lines in tables. If set to `ascii` the characters '-', '|', and '+' for single-width lines and '=' and '#' for double-width lines are used. If set to `unicode` then Unicode box drawing characters will be used. The default is `unicode` if the locale's character encoding is "UTF-8" or `ascii` otherwise.

`-O emphasis={none|bold|underline}`

> How to emphasize text. Bold and underline emphasis are achieved with overstriking, which may not be supported by all the software to which you might pass the output. Default: `none`.

3.4 HTML Output Options

To produce output in HTML format, specify `-o file` on the PSPP command line, optionally followed by any of the options shown in the table below to customize the output format.

`-O format=html`

> Specify the output format. This is only necessary if the file name given on `-o` does not end in `.html`.

`-O charts={template.png|none}`

> Sets the name used for chart files. See Section 3.3 [Plain Text Output Options], page 8, for details.

`-O borders=boolean`

> Decorate the tables with borders. If set to false, the tables produced will have no borders. The default value is true.

`-O css=boolean`

> Use cascading style sheets. Cascading style sheets give an improved appearance and can be used to produce pages which fit a certain web site's style. The default value is true.

3.5 OpenDocument Output Options

To produce output as an OpenDocument text (ODT) document, specify -o *file* on the
PSPP command line. If *file* does not end in .odt, you must also specify -O format=odt.

ODT support is only available if your installation of PSPP was compiled with the libxml2
library.

The OpenDocument output format does not have any configurable options.

3.6 Comma-Separated Value Output Options

To produce output in comma-separated value (CSV) format, specify -o *file* on the PSPP
command line, optionally followed by any of the options shown in the table below to customize the output format.

-O format=csv
> Specify the output format. This is only necessary if the file name given on -o
> does not end in .csv.

-O separator=*field-separator*
> Sets the character used to separate fields. Default: a comma (',').

-O quote=*qualifier*
> Sets *qualifier* as the character used to quote fields that contain white space,
> the separator (or any of the characters in the separator, if it contains more
> than one character), or the quote character itself. If *qualifier* is longer than one
> character, only the first character is used; if *qualifier* is the empty string, then
> fields are never quoted.

-O titles=*boolean*
> Whether table titles (brief descriptions) should be printed. Default: on.

-O captions=*boolean*
> Whether table captions (more extensive descriptions) should be printed. De-
> fault: on.

The CSV format used is an extension to that specified in RFC 4180:

Tables   Each table row is output on a separate line, and each column is output as a
             field. The contents of a cell that spans multiple rows or columns is output only
             for the top-left row and column; the rest are output as empty fields.

Titles   When a table has a title and titles are enabled, the title is output just above
             the table as a single field prefixed by 'Table:'.

Captions   When a table has a caption and captions are enabled, the caption is output
             just below the table as a single field prefixed by 'Caption:'.

Footnotes   Within a table, footnote markers are output as bracketed letters following the
             cell's contents, e.g. '[a]', '[b]', ... The footnotes themselves are output fol-
             lowing the body of the table, as a separate two-column table introduced with a
             line that says 'Footnotes:'. Each row in the table represent one footnote: the
             first column is the marker, the second column is the text.

Text   Text in output is printed as a field on a line by itself. The TITLE and SUBTI-
              TLE produce similar output, prefixed by '`Title:`' or '`Subtitle:`', respectively.

Messages   Errors, warnings, and notes are printed the same way as text.

Charts   Charts are not included in CSV output.

Successive output items are separated by a blank line.

4 Invoking `psppire`

4.1 The graphic user interface

The PSPPIRE graphic user interface for PSPP can perform all functionality of the command line interface. In addition it gives an instantaneous view of the data, variables and statistical output.

The graphic user interface can be started by typing `psppire` at a command prompt. Alternatively many systems have a system of interactive menus or buttons from which `psppire` can be started by a series of mouse clicks.

Once the principles of the PSPP system are understood, the graphic user interface is designed to be largely intuitive, and for this reason is covered only very briefly by this manual.

5 Using PSPP

PSPP is a tool for the statistical analysis of sampled data. You can use it to discover patterns in the data, to explain differences in one subset of data in terms of another subset and to find out whether certain beliefs about the data are justified. This chapter does not attempt to introduce the theory behind the statistical analysis, but it shows how such analysis can be performed using PSPP.

For the purposes of this tutorial, it is assumed that you are using PSPP in its interactive mode from the command line. However, the example commands can also be typed into a file and executed in a post-hoc mode by typing 'pspp *filename*' at a shell prompt, where *filename* is the name of the file containing the commands. Alternatively, from the graphical interface, you can select File → New → Syntax to open a new syntax window and use the Run menu when a syntax fragment is ready to be executed. Whichever method you choose, the syntax is identical.

When using the interactive method, PSPP tells you that it's waiting for your data with a string like PSPP> or data>. In the examples of this chapter, whenever you see text like this, it indicates the prompt displayed by PSPP, *not* something that you should type.

Throughout this chapter reference is made to a number of sample data files. So that you can try the examples for yourself, you should have received these files along with your copy of PSPP.[1]

> **Please note:** Normally these files are installed in the directory /usr/local/share/pspp/examples. If however your system administrator or operating system vendor has chosen to install them in a different location, you will have to adjust the examples accordingly.

5.1 Preparation of Data Files

Before analysis can commence, the data must be loaded into PSPP and arranged such that both PSPP and humans can understand what the data represents. There are two aspects of data:

- The variables — these are the parameters of a quantity which has been measured or estimated in some way. For example height, weight and geographic location are all variables.

- The observations (also called 'cases') of the variables — each observation represents an instance when the variables were measured or observed.

For example, a data set which has the variables *height*, *weight*, and *name*, might have the observations:

```
1881 89.2 Ahmed
1192 107.01 Frank
1230 67 Julie
```

The following sections explain how to define a dataset.

---

[1] These files contain purely fictitious data. They should not be used for research purposes.

5.1.1 Defining Variables

Variables come in two basic types, *viz*: *numeric* and *string*. Variables such as age, height and satisfaction are numeric, whereas name is a string variable. String variables are best reserved for commentary data to assist the human observer. However they can also be used for nominal or categorical data.

Example 5.1 defines two variables *forename* and *height*, and reads data into them by manual input.

```
PSPP> data list list /forename (A12) height.
PSPP> begin data.
data> Ahmed 188
data> Bertram 167
data> Catherine 134.231
data> David 109.1
data> end data
PSPP>
```

Example 5.1: Manual entry of data using the **DATA LIST** command. Two variables *forename* and *height* are defined and subsequently filled with manually entered data.

There are several things to note about this example.

- The words 'data list list' are an example of the **DATA LIST** command. See Section 8.5 [DATA LIST], page 66. It tells PSPP to prepare for reading data. The word 'list' intentionally appears twice. The first occurrence is part of the **DATA LIST** call, whilst the second tells PSPP that the data is to be read as free format data with one record per line.

- The '/' character is important. It marks the start of the list of variables which you wish to define.

- The text 'forename' is the name of the first variable, and '(A12)' says that the variable *forename* is a string variable and that its maximum length is 12 bytes. The second variable's name is specified by the text 'height'. Since no format is given, this variable has the default format. Normally the default format expects numeric data, which should be entered in the locale of the operating system. Thus, the example is correct for English locales and other locales which use a period ('.') as the decimal separator. However if you are using a system with a locale which uses the comma (',') as the decimal separator, then you should in the subsequent lines substitute '.' with ','. Alternatively, you could explicitly tell PSPP that the *height* variable is to be read using a period as its decimal separator by appending the text 'DOT8.3' after the word 'height'. For more information on data formats, see Section 6.7.4 [Input and Output Formats], page 34.

- Normally, PSPP displays the prompt **PSPP>** whenever it's expecting a command. However, when it's expecting data, the prompt changes to **data>** so that you know to enter data and not a command.

- At the end of every command there is a terminating '.' which tells PSPP that the end of a command has been encountered. You should not enter '.' when data is expected (*ie.* when the `data>` prompt is current) since it is appropriate only for terminating commands.

5.1.2 Listing the data

Once the data has been entered, you could type

```
PSPP> list /format=numbered.
```

to list the data. The optional text '`/format=numbered`' requests the case numbers to be shown along with the data. It should show the following output:

```
Case#      forename   height
-----  ------------  --------
    1  Ahmed           188.00
    2  Bertram         167.00
    3  Catherine       134.23
    4  David           109.10
```

Note that the numeric variable *height* is displayed to 2 decimal places, because the format for that variable is '`F8.2`'. For a complete description of the `LIST` command, see Section 8.10 [LIST], page 76.

5.1.3 Reading data from a text file

The previous example showed how to define a set of variables and to manually enter the data for those variables. Manual entering of data is tedious work, and often a file containing the data will be have been previously prepared. Let us assume that you have a file called `mydata.dat` containing the ascii encoded data:

```
Ahmed           188.00
Bertram         167.00
Catherine       134.23
David           109.10
                   .

                   .

                   .
Zachariah       113.02
```

You can can tell the `DATA LIST` command to read the data directly from this file instead of by manual entry, with a command like:

```
PSPP> data list file='mydata.dat' list /forename (A12) height.
```

Notice however, that it is still necessary to specify the names of the variables and their formats, since this information is not contained in the file. It is also possible to specify the file's character encoding and other parameters. For full details refer to see Section 8.5 [DATA LIST], page 66.

5.1.4 Reading data from a pre-prepared PSPP file

When working with other PSPP users, or users of other software which uses the PSPP data format, you may be given the data in a pre-prepared PSPP file. Such files contain not only the data, but the variable definitions, along with their formats, labels and other meta-data. Conventionally, these files (sometimes called "system" files) have the suffix `.sav`, but that is not mandatory. The following syntax loads a file called `my-file.sav`.

```
PSPP> get file='my-file.sav'.
```

You will encounter several instances of this in future examples.

5.1.5 Saving data to a PSPP file.

If you want to save your data, along with the variable definitions so that you or other PSPP users can use it later, you can do this with the **SAVE** command.

The following syntax will save the existing data and variables to a file called `my-new-file.sav`.

```
PSPP> save outfile='my-new-file.sav'.
```

If `my-new-file.sav` already exists, then it will be overwritten. Otherwise it will be created.

5.1.6 Reading data from other sources

Sometimes it's useful to be able to read data from comma separated text, from spreadsheets, databases or other sources. In these instances you should use the **GET DATA** command (see Section 9.4 [GET DATA], page 83).

5.1.7 Exiting PSPP

Use the **FINISH** command to exit PSPP:

```
PSPP> finish.
```

5.2 Data Screening and Transformation

Once data has been entered, it is often desirable, or even necessary, to transform it in some way before performing analysis upon it. At the very least, it's good practice to check for errors.

5.2.1 Identifying incorrect data

Data from real sources is rarely error free. PSPP has a number of procedures which can be used to help identify data which might be incorrect.

The **DESCRIPTIVES** command (see Section 15.1 [DESCRIPTIVES], page 127) is used to generate simple linear statistics for a dataset. It is also useful for identifying potential problems in the data. The example file `physiology.sav` contains a number of physiological measurements of a sample of healthy adults selected at random. However, the data entry clerk made a number of mistakes when entering the data. Example 5.2 illustrates the use of **DESCRIPTIVES** to screen this data and identify the erroneous values.

Chapter 5: Using PSPP

```
PSPP> get file='/usr/local/share/pspp/examples/physiology.sav'.
PSPP> descriptives sex, weight, height.
   Output:
DESCRIPTIVES.  Valid cases = 40; cases with missing value(s) = 0.
+--------#--+-------+-------+-------+-------+
|Variable# N|  Mean |Std Dev|Minimum|Maximum|
#========#==#=======#=======#=======#=======#
|sex     #40|   .45|    .50|    .00|   1.00|
|height  #40|1677.12| 262.87| 179.00|1903.00|
|weight  #40|  72.12|  26.70| -55.60|  92.07|
+--------#--+-------+-------+-------+-------+
```

Example 5.2: Using the DESCRIPTIVES command to display simple summary information about the data. In this case, the results show unexpectedly low values in the Minimum column, suggesting incorrect data entry.

In the output of Example 5.2, the most interesting column is the minimum value. The *weight* variable has a minimum value of less than zero, which is clearly erroneous. Similarly, the *height* variable's minimum value seems to be very low. In fact, it is more than 5 standard deviations from the mean, and is a seemingly bizarre height for an adult person. We can examine the data in more detail with the EXAMINE command (see Section 15.3 [EXAMINE], page 130):

In Example 5.3 you can see that the lowest value of *height* is 179 (which we suspect to be erroneous), but the second lowest is 1598 which we know from the DESCRIPTIVES command is within 1 standard deviation from the mean. Similarly the *weight* variable has a lowest value which is negative but a plausible value for the second lowest value. This suggests that the two extreme values are outliers and probably represent data entry errors.

```
    [... continue from Example 5.2]
PSPP> examine height, weight /statistics=extreme(3).
    Output:
#===============================#==========#======#
#                               #Case Number| Value #
#===============================#==========#======#
#Height in millimetres Highest 1#       14|1903.00#
#                             2#       15|1884.00#
#                             3#       12|1801.65#
#                 ----------#-----------+-------#
#                     Lowest 1#       30| 179.00#
#                             2#       31|1598.00#
#                             3#       28|1601.00#
#                 ----------#-----------+-------#
#Weight in kilograms   Highest 1#       13|  92.07#
#                             2#        5|  92.07#
#                             3#       17|  91.74#
#                 ----------#-----------+-------#
#                     Lowest 1#       38| -55.60#
#                             2#       39|  54.48#
#                             3#       33|  55.45#
#===============================#==========#======#
```

Example 5.3: Using the EXAMINE command to see the extremities of the data for different variables. Cases 30 and 38 seem to contain values very much lower than the rest of the data. They are possibly erroneous.

5.2.2 Dealing with suspicious data

If possible, suspect data should be checked and re-measured. However, this may not always be feasible, in which case the researcher may decide to disregard these values. PSPP has a feature whereby data can assume the special value 'SYSMIS', and will be disregarded in future analysis. See Section 6.6 [Missing Observations], page 32. You can set the two suspect values to the 'SYSMIS' value using the RECODE command.

```
PSPP
> recode height (179 = SYSMIS).
PSPP
> recode weight (LOWEST THRU 0 = SYSMIS).
```

The first command says that for any observation which has a *height* value of 179, that value should be changed to the SYSMIS value. The second command says that any *weight* values of zero or less should be changed to SYSMIS. From now on, they will be ignored in analysis. For detailed information about the RECODE command see Section 12.7 [RECODE], page 116.

If you now re-run the DESCRIPTIVES or EXAMINE commands in Example 5.2 and Example 5.3 you will see a data summary with more plausible parameters. You will also notice that the data summaries indicate the two missing values.

5.2.3 Inverting negatively coded variables

Data entry errors are not the only reason for wanting to recode data. The sample file `hotel.sav` comprises data gathered from a customer satisfaction survey of clients at a particular hotel. In Example 5.4, this file is loaded for analysis. The line `display dictionary.` tells PSPP to display the variables and associated data. The output from this command has been omitted from the example for the sake of clarity, but you will notice that each of the variables *v1*, *v2* ... *v5* are measured on a 5 point Likert scale, with 1 meaning "Strongly disagree" and 5 meaning "Strongly agree". Whilst variables *v1*, *v2* and *v4* record responses to a positively posed question, variables *v3* and *v5* are responses to negatively worded questions. In order to perform meaningful analysis, we need to recode the variables so that they all measure in the same direction. We could use the RECODE command, with syntax such as:

```
recode v3 (1 = 5) (2 = 4) (4 = 2) (5 = 1).
```

However an easier and more elegant way uses the COMPUTE command (see Section 12.3 [COMPUTE], page 113). Since the variables are Likert variables in the range (1 ... 5), subtracting their value from 6 has the effect of inverting them:

```
compute var = 6 - var.
```

Example 5.4 uses this technique to recode the variables *v3* and *v5*. After applying COMPUTE for both variables, all subsequent commands will use the inverted values.

5.2.4 Testing data consistency

A sensible check to perform on survey data is the calculation of reliability. This gives the statistician some confidence that the questionnaires have been completed thoughtfully. If you examine the labels of variables *v1*, *v3* and *v4*, you will notice that they ask very similar questions. One would therefore expect the values of these variables (after recoding) to closely follow one another, and we can test that with the RELIABILITY command (see Section 15.16 [RELIABILITY], page 151). Example 5.4 shows a PSPP session where the user (after recoding negatively scaled variables) requests reliability statistics for *v1*, *v3* and *v4*.

```
PSPP> get file='/usr/local/share/pspp/examples/hotel.sav'.
PSPP> display dictionary.
PSPP> * recode negatively worded questions.
PSPP> compute v3 = 6 - v3.
PSPP> compute v5 = 6 - v5.
PSPP> reliability v1, v3, v4.

    Output (dictionary information omitted for clarity):

1.1 RELIABILITY.  Case Processing Summary
#==============#==#======#
#              # N|   %  #
#==============#==#======#
#Cases Valid   #17|100.00#
#      Excluded# 0|   .00#
#      Total   #17|100.00#
#==============#==#======#


1.2 RELIABILITY.  Reliability Statistics
#=================#==========#
#Cronbach's Alpha#N of Items#
#=================#==========#
#             .81#         3#
#=================#==========#
```

Example 5.4: Recoding negatively scaled variables, and testing for reliability with the
RELIABILITY command. The Cronbach Alpha coefficient suggests a high degree of reliability
among variables *v1*, *v3* and *v4*.

As a rule of thumb, many statisticians consider a value of Cronbach's Alpha of 0.7 or
higher to indicate reliable data. Here, the value is 0.81 so the data and the recoding that
we performed are vindicated.

5.2.5 Testing for normality

Many statistical tests rely upon certain properties of the data. One common property, upon
which many linear tests depend, is that of normality — the data must have been drawn
from a normal distribution. It is necessary then to ensure normality before deciding upon
the test procedure to use. One way to do this uses the EXAMINE command.

In Example 5.5, a researcher was examining the failure rates of equipment produced by
an engineering company. The file repairs.sav contains the mean time between failures
(*mtbf*) of some items of equipment subject to the study. Before performing linear analysis
on the data, the researcher wanted to ascertain that the data is normally distributed.

A normal distribution has a skewness and kurtosis of zero. Looking at the skewness
of *mtbf* in Example 5.5 it is clear that the mtbf figures have a lot of positive skew and
are therefore not drawn from a normally distributed variable. Positive skew can often be

compensated for by applying a logarithmic transformation. This is done with the COMPUTE
command in the line

```
compute mtbf_ln = ln (mtbf).
```

Rather than redefining the existing variable, this use of COMPUTE defines a new variable
mtbf_ln which is the natural logarithm of *mtbf*. The final command in this example calls
EXAMINE on this new variable, and it can be seen from the results that both the skewness
and kurtosis for *mtbf_ln* are very close to zero. This provides some confidence that the
mtbf_ln variable is normally distributed and thus safe for linear analysis. In the event that
no suitable transformation can be found, then it would be worth considering an appropriate
non-parametric test instead of a linear one. See Section 15.10 [NPAR TESTS], page 141,
for information about non-parametric tests.

```
PSPP> get file='/usr/local/share/pspp/examples/repairs.sav'.
PSPP> examine mtbf
            /statistics=descriptives.
PSPP> compute mtbf_ln = ln (mtbf).
PSPP> examine mtbf_ln
            /statistics=descriptives.
```

  Output:

1.2 EXAMINE. Descriptives

		Statistic	Std. Error
mtbf	Mean	8.32	1.62
	95% Confidence Interval for Mean Lower Bound	4.85	
	Upper Bound	11.79	
	5% Trimmed Mean	7.69	
	Median	8.12	
	Variance	39.21	
	Std. Deviation	6.26	
	Minimum	1.63	
	Maximum	26.47	
	Range	24.84	
	Interquartile Range	5.83	
	Skewness	1.85	.58
	Kurtosis	4.49	1.12

2.2 EXAMINE. Descriptives

		Statistic	Std. Error
mtbf_ln	Mean	1.88	.19
	95% Confidence Interval for Mean Lower Bound	1.47	
	Upper Bound	2.29	
	5% Trimmed Mean	1.88	
	Median	2.09	
	Variance	.54	
	Std. Deviation	.74	
	Minimum	.49	
	Maximum	3.28	
	Range	2.79	
	Interquartile Range	.92	
	Skewness	-.16	.58
	Kurtosis	-.09	1.12

Example 5.5: Testing for normality using the EXAMINE command and applying a logarithmic transformation. The *mtbf* variable has a large positive skew and is therefore unsuitable for linear statistical analysis. However the transformed variable (*mtbf_ln*) is close to normal and would appear to be more suitable.

5.3 Hypothesis Testing

One of the most fundamental purposes of statistical analysis is hypothesis testing. Researchers commonly need to test hypotheses about a set of data. For example, she might want to test whether one set of data comes from the same distribution as another, or whether the mean of a dataset significantly differs from a particular value. This section presents just some of the possible tests that PSPP offers.

The researcher starts by making a *null hypothesis*. Often this is a hypothesis which he suspects to be false. For example, if he suspects that A is greater than B he will state the null hypothesis as $A = B$.[2]

The *p-value* is a recurring concept in hypothesis testing. It is the highest acceptable probability that the evidence implying a null hypothesis is false, could have been obtained when the null hypothesis is in fact true. Note that this is not the same as "the probability of making an error" nor is it the same as "the probability of rejecting a hypothesis when it is true".

5.3.1 Testing for differences of means

A common statistical test involves hypotheses about means. The T-TEST command is used to find out whether or not two separate subsets have the same mean.

Example 5.6 uses the file `physiology.sav` previously encountered. A researcher suspected that the heights and core body temperature of persons might be different depending upon their sex. To investigate this, he posed two null hypotheses:

- The mean heights of males and females in the population are equal.
- The mean body temperature of males and females in the population are equal.

For the purposes of the investigation the researcher decided to use a p-value of 0.05.

In addition to the T-test, the T-TEST command also performs the Levene test for equal variances. If the variances are equal, then a more powerful form of the T-test can be used. However if it is unsafe to assume equal variances, then an alternative calculation is necessary. PSPP performs both calculations.

For the *height* variable, the output shows the significance of the Levene test to be 0.33 which means there is a 33% probability that the Levene test produces this outcome when the variances are equal. Had the significance been less than 0.05, then it would have been unsafe to assume that the variances were equal. However, because the value is higher than 0.05 the homogeneity of variances assumption is safe and the "Equal Variances" row (the more powerful test) can be used. Examining this row, the two tailed significance for the *height* t-test is less than 0.05, so it is safe to reject the null hypothesis and conclude that the mean heights of males and females are unequal.

For the *temperature* variable, the significance of the Levene test is 0.58 so again, it is safe to use the row for equal variances. The equal variances row indicates that the two tailed significance for *temperature* is 0.20. Since this is greater than 0.05 we must reject the null hypothesis and conclude that there is insufficient evidence to suggest that the body temperature of male and female persons are different.

---

[2] This example assumes that it is already proven that B is not greater than A.

```
PSPP> get file='/usr/local/share/pspp/examples/physiology.sav'.
PSPP> recode height (179 = SYSMIS).
PSPP> t-test group=sex(0,1) /variables = height temperature.
   Output:
1.1 T-TEST.  Group Statistics
#================#==#======#==============#=======#
#            sex | N|  Mean |Std. Deviation|SE. Mean#
#================#==#======#==============#=======#
#height    Male  |22|1796.49|        49.71|  10.60#
#          Female|17|1610.77|        25.43|   6.17#
#temperature Male  |22|  36.68|         1.95|    .42#
#          Female|18|  37.43|         1.61|    .38#
#================#==#======#==============#=======#
1.2 T-TEST.  Independent Samples Test
#========================#========#========================= =#
#                        # Levene's| t-test for Equality of Means     #
#                        #----+----+------+-----+------+---------+- -#
#                        #    |    |      |     |      |         |  #
#                        #    |    |      |     |Sig. 2|         |  #
#                        # F |Sig.|  t  |  df |tailed|Mean Diff|  #
#========================#====#====#======#=====#======#=========#= =#
#height     Equal variances# .97|  .33| 14.02|37.00|   .00|   185.72| ... #
#          Unequal variances#   |    | 15.15|32.71|   .00|   185.72| ... #
#temperature Equal variances# .31|  .58| -1.31|38.00|   .20|    -.75| ... #
#          Unequal variances#   |    | -1.33|37.99|   .19|    -.75| ... #
#========================#====#====#======#=====#======#=========#= =#
```

Example 5.6: The T-TEST command tests for differences of means. Here, the *height* variable's two tailed significance is less than 0.05, so the null hypothesis can be rejected. Thus, the evidence suggests there is a difference between the heights of male and female persons. However the significance of the test for the *temperature* variable is greater than 0.05 so the null hypothesis cannot be rejected, and there is insufficient evidence to suggest a difference in body temperature.

5.3.2 Linear Regression

Linear regression is a technique used to investigate if and how a variable is linearly related to others. If a variable is found to be linearly related, then this can be used to predict future values of that variable.

In example Example 5.7, the service department of the company wanted to be able to predict the time to repair equipment, in order to improve the accuracy of their quotations. It was suggested that the time to repair might be related to the time between failures and the duty cycle of the equipment. The p-value of 0.1 was chosen for this investigation. In order to investigate this hypothesis, the REGRESSION command was used. This command not only tests if the variables are related, but also identifies the potential linear relationship. See Section 15.15 [REGRESSION], page 149.

```
PSPP> get file='/usr/local/share/pspp/examples/repairs.sav'.
PSPP> regression /variables = mtbf duty_cycle /dependent = mttr.
PSPP> regression /variables = mtbf /dependent = mttr.
   Output:
1.3(1) REGRESSION.  Coefficients
#=================================================#====#=========#====#=====#
#                                                 #  B |Std. Error|Beta|  t  #
#=======#=========================================#====#=========#====#=====#
#       |(Constant)                               #9.81|     1.50|  .00| 6.54#
#       |Mean time between failures (months)      #3.10|      .10|  .99|32.43#
#       |Ratio of working to non-working time     #1.09|     1.78|  .02|  .61#
#       |                                         #    |         |    |     #
#=======#=========================================#====#=========#====#=====#


1.3(2) REGRESSION.  Coefficients
#=================================================#===========#
#                                                 #Significance#
#=======#=========================================#===========#
#       |(Constant)                               #       .10#
#       |Mean time between failures (months)       #       .00#
#       |Ratio of working to non-working time      #       .55#
#       |                                         #         #
#=======#=========================================#===========#
2.3(1) REGRESSION.  Coefficients
#=================================================#=====#=========#====#=====#
#                                                 #  B  |Std. Error|Beta|  t  #
#=======#=========================================#=====#=========#====#=====#
#       |(Constant)                               #10.50|      .96|  .00|10.96#
#       |Mean time between failures (months)      # 3.11|      .09|  .99|33.39#
#       |                                         #     |         |    |     #
#=======#=========================================#=====#=========#====#=====#


2.3(2) REGRESSION.  Coefficients
#=================================================#===========#
#                                                 #Significance#
#=======#=========================================#===========#
#       |(Constant)                               #       .06#
#       |Mean time between failures (months)      #       .00#
#       |                                         #         #
#=======#=========================================#===========#
```

Example 5.7: Linear regression analysis to find a predictor for *mttr*. The first attempt, including *duty_cycle*, produces some unacceptable high significance values. However the second attempt, which excludes *duty_cycle*, produces significance values no higher than 0.06. This suggests that *mtbf* alone may be a suitable predictor for *mttr*.

The coefficients in the first table suggest that the formula $mttr = 9.81 + 3.1 \times mtbf + 1.09 \times duty\_cycle$ can be used to predict the time to repair. However, the significance value for the $duty\_cycle$ coefficient is very high, which would make this an unsafe predictor. For this reason, the test was repeated, but omitting the $duty\_cycle$ variable. This time, the significance of all coefficients no higher than 0.06, suggesting that at the 0.06 level, the formula $mttr = 10.5 + 3.11 \times mtbf$ is a reliable predictor of the time to repair.

6 The PSPP language

This chapter discusses elements common to many PSPP commands. Later chapters will describe individual commands in detail.

6.1 Tokens

PSPP divides most syntax file lines into series of short chunks called *tokens*. Tokens are then grouped to form commands, each of which tells PSPP to take some action—read in data, write out data, perform a statistical procedure, etc. Each type of token is described below.

Identifiers Identifiers are names that typically specify variables, commands, or subcommands. The first character in an identifier must be a letter, '#', or '@'. The remaining characters in the identifier must be letters, digits, or one of the following special characters:

. _ $ # @

Identifiers may be any length, but only the first 64 bytes are significant. Identifiers are not case-sensitive: `foobar`, `Foobar`, `FooBar`, `FOOBAR`, and `FoObaR` are different representations of the same identifier.

Some identifiers are reserved. Reserved identifiers may not be used in any context besides those explicitly described in this manual. The reserved identifiers are:

ALL AND BY EQ GE GT LE LT NE NOT OR TO WITH

Keywords Keywords are a subclass of identifiers that form a fixed part of command syntax. For example, command and subcommand names are keywords. Keywords may be abbreviated to their first 3 characters if this abbreviation is unambiguous. (Unique abbreviations of 3 or more characters are also accepted: 'FRE', 'FREQ', and 'FREQUENCIES' are equivalent when the last is a keyword.)

Reserved identifiers are always used as keywords. Other identifiers may be used both as keywords and as user-defined identifiers, such as variable names.

Numbers Numbers are expressed in decimal. A decimal point is optional. Numbers may be expressed in scientific notation by adding 'e' and a base-10 exponent, so that '1.234e3' has the value 1234. Here are some more examples of valid numbers:

-5 3.14159265359 1e100 -.707 8945.

Negative numbers are expressed with a '-' prefix. However, in situations where a literal '-' token is expected, what appears to be a negative number is treated as '-' followed by a positive number.

No white space is allowed within a number token, except for horizontal white space between '-' and the rest of the number.

The last example above, '8945.' will be interpreted as two tokens, '8945' and '.', if it is the last token on a line. See Section 6.2 [Forming commands of tokens], page 29.

Strings Strings are literal sequences of characters enclosed in pairs of single quotes (' ') or double quotes (" "). To include the character used for quoting in the string, double it, e.g. ''it''s an apostrophe''. White space and case of letters are significant inside strings.

Strings can be concatenated using '+', so that '"a" + 'b' + 'c'' is equivalent to ''abc''. So that a long string may be broken across lines, a line break may precede or follow, or both precede and follow, the '+'. (However, an entirely blank line preceding or following the '+' is interpreted as ending the current command.)

Strings may also be expressed as hexadecimal character values by prefixing the initial quote character by 'x' or 'X'. Regardless of the syntax file or active dataset's encoding, the hexadecimal digits in the string are interpreted as Unicode characters in UTF-8 encoding.

Individual Unicode code points may also be expressed by specifying the hexadecimal code point number in single or double quotes preceded by 'u' or 'U'. For example, Unicode code point U+1D11E, the musical G clef character, could be expressed as U'1D11E'. Invalid Unicode code points (above U+10FFFF or in between U+D800 and U+DFFF) are not allowed.

When strings are concatenated with '+', each segment's prefix is considered individually. For example, 'The G clef symbol is:' + u"1d11e" + "." inserts a G clef symbol in the middle of an otherwise plain text string.

Punctuators and Operators

These tokens are the punctuators and operators:

, / = ( ) + - * / ** < <= <> > >= ~= & | .

Most of these appear within the syntax of commands, but the period ('.') punctuator is used only at the end of a command. It is a punctuator only as the last character on a line (except white space). When it is the last non-space character on a line, a period is not treated as part of another token, even if it would otherwise be part of, e.g., an identifier or a floating-point number.

6.2 Forming commands of tokens

Most PSPP commands share a common structure. A command begins with a command name, such as FREQUENCIES, DATA LIST, or N OF CASES. The command name may be abbreviated to its first word, and each word in the command name may be abbreviated to its first three or more characters, where these abbreviations are unambiguous.

The command name may be followed by one or more *subcommands*. Each subcommand begins with a subcommand name, which may be abbreviated to its first three letters. Some subcommands accept a series of one or more specifications, which follow the subcommand name, optionally separated from it by an equals sign ('='). Specifications may be separated from each other by commas or spaces. Each subcommand must be separated from the next (if any) by a forward slash ('/').

There are multiple ways to mark the end of a command. The most common way is to end the last line of the command with a period ('.') as described in the previous section (see Section 6.1 [Tokens], page 28). A blank line, or one that consists only of white space or comments, also ends a command.

6.3 Syntax Variants

There are three variants of command syntax, which vary only in how they detect the end of one command and the start of the next.

In *interactive mode*, which is the default for syntax typed at a command prompt, a period as the last non-blank character on a line ends a command. A blank line also ends a command.

In *batch mode*, an end-of-line period or a blank line also ends a command. Additionally, it treats any line that has a non-blank character in the leftmost column as beginning a new command. Thus, in batch mode the second and subsequent lines in a command must be indented.

Regardless of the syntax mode, a plus sign, minus sign, or period in the leftmost column of a line is ignored and causes that line to begin a new command. This is most useful in batch mode, in which the first line of a new command could not otherwise be indented, but it is accepted regardless of syntax mode.

The default mode for reading commands from a file is *auto mode*. It is the same as batch mode, except that a line with a non-blank in the leftmost column only starts a new command if that line begins with the name of a PSPP command. This correctly interprets most valid PSPP syntax files regardless of the syntax mode for which they are intended.

The `--interactive` (or `-i`) or `--batch` (or `-b`) options set the syntax mode for files listed on the PSPP command line. See Section 3.1 [Main Options], page 4, for more details.

6.4 Types of Commands

Commands in PSPP are divided roughly into six categories:

Utility commands

> Set or display various global options that affect PSPP operations. May appear anywhere in a syntax file. See Chapter 16 [Utility commands], page 153.

File definition commands

> Give instructions for reading data from text files or from special binary "system files". Most of these commands replace any previous data or variables with new data or variables. At least one file definition command must appear before the first command in any of the categories below. See Chapter 8 [Data Input and Output], page 64.

Input program commands

> Though rarely used, these provide tools for reading data files in arbitrary textual or binary formats. See Section 8.9 [INPUT PROGRAM], page 73.

Transformations

> Perform operations on data and write data to output files. Transformations are not carried out until a procedure is executed.

Restricted transformations

> Transformations that cannot appear in certain contexts. See Section 6.5 [Order of Commands], page 31, for details.

Procedures

Analyze data, writing results of analyses to the listing file. Cause transformations specified earlier in the file to be performed. In a more general sense, a *procedure* is any command that causes the active dataset (the data) to be read.

6.5 Order of Commands

PSPP does not place many restrictions on ordering of commands. The main restriction is that variables must be defined before they are otherwise referenced. This section describes the details of command ordering, but most users will have no need to refer to them.

PSPP possesses five internal states, called *initial*, *input-program* *file-type*, *transformation*, and *procedure* states. (Please note the distinction between the INPUT PROGRAM and FILE TYPE *commands* and the *input-program* and *file-type states*.)

PSPP starts in the initial state. Each successful completion of a command may cause a state transition. Each type of command has its own rules for state transitions:

Utility commands

- Valid in any state.
- Do not cause state transitions. Exception: when N OF CASES is executed in the procedure state, it causes a transition to the transformation state.

DATA LIST

- Valid in any state.
- When executed in the initial or procedure state, causes a transition to the transformation state.
- Clears the active dataset if executed in the procedure or transformation state.

INPUT PROGRAM

- Invalid in input-program and file-type states.
- Causes a transition to the intput-program state.
- Clears the active dataset.

FILE TYPE

- Invalid in intput-program and file-type states.
- Causes a transition to the file-type state.
- Clears the active dataset.

Other file definition commands

- Invalid in input-program and file-type states.
- Cause a transition to the transformation state.
- Clear the active dataset, except for ADD FILES, MATCH FILES, and UPDATE.

Transformations

- Invalid in initial and file-type states.
- Cause a transition to the transformation state.

Restricted transformations

- Invalid in initial, input-program, and file-type states.

- Cause a transition to the transformation state.

Procedures

- Invalid in initial, input-program, and file-type states.
- Cause a transition to the procedure state.

6.6 Handling missing observations

PSPP includes special support for unknown numeric data values. Missing observations are assigned a special value, called the *system-missing value*. This "value" actually indicates the absence of a value; it means that the actual value is unknown. Procedures automatically exclude from analyses those observations or cases that have missing values. Details of missing value exclusion depend on the procedure and can often be controlled by the user; refer to descriptions of individual procedures for details.

The system-missing value exists only for numeric variables. String variables always have a defined value, even if it is only a string of spaces.

Variables, whether numeric or string, can have designated *user-missing values*. Every user-missing value is an actual value for that variable. However, most of the time user-missing values are treated in the same way as the system-missing value.

For more information on missing values, see the following sections: Section 6.7 [Datasets], page 32, Section 11.6 [MISSING VALUES], page 102, Chapter 7 [Expressions], page 46. See also the documentation on individual procedures for information on how they handle missing values.

6.7 Datasets

PSPP works with data organized into *datasets*. A dataset consists of a set of *variables*, which taken together are said to form a *dictionary*, and one or more *cases*, each of which has one value for each variable.

At any given time PSPP has exactly one distinguished dataset, called the *active dataset*. Most PSPP commands work only with the active dataset. In addition to the active dataset, PSPP also supports any number of additional open datasets. The `DATASET` commands can choose a new active dataset from among those that are open, as well as create and destroy datasets (see Section 8.4 [DATASET], page 65).

The sections below describe variables in more detail.

6.7.1 Attributes of Variables

Each variable has a number of attributes, including:

Name   An identifier, up to 64 bytes long. Each variable must have a different name. See Section 6.1 [Tokens], page 28.

Some system variable names begin with '`$`', but user-defined variables' names may not begin with '`$`'.

The final character in a variable name should not be '`.`', because such an identifier will be misinterpreted when it is the final token on a line: `FOO.` will be divided into two separate tokens, '`FOO`' and '`.`', indicating end-of-command. See Section 6.1 [Tokens], page 28.

The final character in a variable name should not be '_', because some such identifiers are used for special purposes by PSPP procedures.

As with all PSPP identifiers, variable names are not case-sensitive. PSPP capitalizes variable names on output the same way they were capitalized at their point of definition in the input.

Type   Numeric or string.

Width   (string variables only) String variables with a width of 8 characters or fewer are called *short string variables*. Short string variables may be used in a few contexts where *long string variables* (those with widths greater than 8) are not allowed.

Position Variables in the dictionary are arranged in a specific order. **DISPLAY** can be used to show this order: see Section 11.3 [DISPLAY], page 100.

Initialization

Either reinitialized to 0 or spaces for each case, or left at its existing value. See Section 11.5 [LEAVE], page 101.

Missing values

Optionally, up to three values, or a range of values, or a specific value plus a range, can be specified as *user-missing values*. There is also a *system-missing value* that is assigned to an observation when there is no other obvious value for that observation. Observations with missing values are automatically excluded from analyses. User-missing values are actual data values, while the system-missing value is not a value at all. See Section 6.6 [Missing Observations], page 32.

Variable label

A string that describes the variable. See Section 11.15 [VARIABLE LABELS], page 107.

Value label

Optionally, these associate each possible value of the variable with a string. See Section 11.12 [VALUE LABELS], page 105.

Print format

Display width, format, and (for numeric variables) number of decimal places. This attribute does not affect how data are stored, just how they are displayed. Example: a width of 8, with 2 decimal places. See Section 6.7.4 [Input and Output Formats], page 34.

Write format

Similar to print format, but used by the **WRITE** command (see Section 8.17 [WRITE], page 80).

Custom attributes

User-defined associations between names and values. See Section 11.14 [VARIABLE ATTRIBUTE], page 106.

Role   The intended role of a variable for use in dialog boxes in graphical user interfaces. See Section 11.19 [VARIABLE ROLE], page 108.

6.7.2 Variables Automatically Defined by PSPP

There are seven system variables. These are not like ordinary variables because system variables are not always stored. They can be used only in expressions. These system variables, whose values and output formats cannot be modified, are described below.

$CASENUM   Case number of the case at the moment. This changes as cases are shuffled around.

$DATE   Date the PSPP process was started, in format A9, following the pattern DD MMM YY.

$JDATE   Number of days between 15 Oct 1582 and the time the PSPP process was started.

$LENGTH   Page length, in lines, in format F11.

$SYSMIS   System missing value, in format F1.

$TIME   Number of seconds between midnight 14 Oct 1582 and the time the active dataset was read, in format F20.

$WIDTH   Page width, in characters, in format F3.

6.7.3 Lists of variable names

To refer to a set of variables, list their names one after another. Optionally, their names may be separated by commas. To include a range of variables from the dictionary in the list, write the name of the first and last variable in the range, separated by TO. For instance, if the dictionary contains six variables with the names ID, X1, X2, GOAL, MET, and NEXTGOAL, in that order, then X2 TO MET would include variables X2, GOAL, and MET.

Commands that define variables, such as DATA LIST, give TO an alternate meaning. With these commands, TO define sequences of variables whose names end in consecutive integers. The syntax is two identifiers that begin with the same root and end with numbers, separated by TO. The syntax X1 TO X5 defines 5 variables, named X1, X2, X3, X4, and X5. The syntax ITEM0008 TO ITEM0013 defines 6 variables, named ITEM0008, ITEM0009, ITEM0010, ITEM0011, ITEM0012, and ITEM00013. The syntaxes QUES001 TO QUES9 and QUES6 TO QUES3 are invalid.

After a set of variables has been defined with DATA LIST or another command with this method, the same set can be referenced on later commands using the same syntax.

6.7.4 Input and Output Formats

An *input format* describes how to interpret the contents of an input field as a number or a string. It might specify that the field contains an ordinary decimal number, a time or date, a number in binary or hexadecimal notation, or one of several other notations. Input formats are used by commands such as DATA LIST that read data or syntax files into the PSPP active dataset.

Every input format corresponds to a default *output format* that specifies the formatting used when the value is output later. It is always possible to explicitly specify an output format that resembles the input format. Usually, this is the default, but in cases where the input format is unfriendly to human readability, such as binary or hexadecimal formats, the default output format is an easier-to-read decimal format.

Every variable has two output formats, called its *print format* and *write format*. Print formats are used in most output contexts; write formats are used only by WRITE (see Section 8.17 [WRITE], page 80). Newly created variables have identical print and write formats, and FORMATS, the most commonly used command for changing formats (see Section 11.4 [FORMATS], page 101), sets both of them to the same value as well. Thus, most of the time, the distinction between print and write formats is unimportant.

Input and output formats are specified to PSPP with a *format specification* of the form *TYPE*w or *TYPE*w.d, where *TYPE* is one of the format types described later, w is a field width measured in columns, and d is an optional number of decimal places. If d is omitted, a value of 0 is assumed. Some formats do not allow a nonzero d to be specified.

The following sections describe the input and output formats supported by PSPP.

6.7.4.1 Basic Numeric Formats

The basic numeric formats are used for input and output of real numbers in standard or scientific notation. The following table shows an example of how each format displays positive and negative numbers with the default decimal point setting:

Format	3141.59	−3141.59
F8.2	3141.59	−3141.59
COMMA9.2	3,141.59	−3,141.59
DOT9.2	3.141,59	−3.141,59
DOLLAR10.2	$3,141.59	−$3,141.59
PCT9.2	3141.59%	−3141.59%
E8.1	3.1E+003	−3.1E+003

On output, numbers in F format are expressed in standard decimal notation with the requested number of decimal places. The other formats output some variation on this style:

- Numbers in COMMA format are additionally grouped every three digits by inserting a grouping character. The grouping character is ordinarily a comma, but it can be changed to a period (see [SET DECIMAL], page 159).

- DOT format is like COMMA format, but it interchanges the role of the decimal point and grouping characters. That is, the current grouping character is used as a decimal point and vice versa.

- DOLLAR format is like COMMA format, but it prefixes the number with '$'.

- PCT format is like F format, but adds '%' after the number.

- The E format always produces output in scientific notation.

On input, the basic numeric formats accept positive and numbers in standard decimal notation or scientific notation. Leading and trailing spaces are allowed. An empty or all-spaces field, or one that contains only a single period, is treated as the system missing value.

In scientific notation, the exponent may be introduced by a sign ('+' or '−'), or by one of the letters 'e' or 'd' (in uppercase or lowercase), or by a letter followed by a sign. A single space may follow the letter or the sign or both.

On fixed-format **DATA LIST** (see Section 8.5.1 [DATA LIST FIXED], page 66) and in a few other contexts, decimals are implied when the field does not contain a decimal point. In F6.5 format, for example, the field **314159** is taken as the value 3.14159 with implied decimals. Decimals are never implied if an explicit decimal point is present or if scientific notation is used.

E and F formats accept the basic syntax already described. The other formats allow some additional variations:

- COMMA, DOLLAR, and DOT formats ignore grouping characters within the integer part of the input field. The identity of the grouping character depends on the format.

- DOLLAR format allows a dollar sign to precede the number. In a negative number, the dollar sign may precede or follow the minus sign.

- PCT format allows a percent sign to follow the number.

All of the basic number formats have a maximum field width of 40 and accept no more than 16 decimal places, on both input and output. Some additional restrictions apply:

- As input formats, the basic numeric formats allow no more decimal places than the field width. As output formats, the field width must be greater than the number of decimal places; that is, large enough to allow for a decimal point and the number of requested decimal places. DOLLAR and PCT formats must allow an additional column for '\$' or '%'.

- The default output format for a given input format increases the field width enough to make room for optional input characters. If an input format calls for decimal places, the width is increased by 1 to make room for an implied decimal point. COMMA, DOT, and DOLLAR formats also increase the output width to make room for grouping characters. DOLLAR and PCT further increase the output field width by 1 to make room for '\$' or '%'. The increased output width is capped at 40, the maximum field width.

- The E format is exceptional. For output, E format has a minimum width of 7 plus the number of decimal places. The default output format for an E input format is an E format with at least 3 decimal places and thus a minimum width of 10.

More details of basic numeric output formatting are given below:

- Output rounds to nearest, with ties rounded away from zero. Thus, 2.5 is output as **3** in F1.0 format, and -1.125 as **-1.13** in F5.1 format.

- The system-missing value is output as a period in a field of spaces, placed in the decimal point's position, or in the rightmost column if no decimal places are requested. A period is used even if the decimal point character is a comma.

- A number that does not fill its field is right-justified within the field.

- A number is too large for its field causes decimal places to be dropped to make room. If dropping decimals does not make enough room, scientific notation is used if the field is wide enough. If a number does not fit in the field, even in scientific notation, the overflow is indicated by filling the field with asterisks ('*').

- COMMA, DOT, and DOLLAR formats insert grouping characters only if space is available for all of them. Grouping characters are never inserted when all decimal places must be dropped. Thus, 1234.56 in COMMA5.2 format is output as ' 1235' without a comma, even though there is room for one, because all decimal places were dropped.

- DOLLAR or PCT format drop the '$' or '%' only if the number would not fit at all without it. Scientific notation with '$' or '%' is preferred to ordinary decimal notation without it.

- Except in scientific notation, a decimal point is included only when it is followed by a digit. If the integer part of the number being output is 0, and a decimal point is included, then the zero before the decimal point is dropped.

  In scientific notation, the number always includes a decimal point, even if it is not followed by a digit.

- A negative number includes a minus sign only in the presence of a nonzero digit: -0.01 is output as '-.01' in F4.2 format but as '   .0' in F4.1 format. Thus, a "negative zero" never includes a minus sign.

- In negative numbers output in DOLLAR format, the dollar sign follows the negative sign. Thus, -9.99 in DOLLAR6.2 format is output as `-$9.99`.

- In scientific notation, the exponent is output as 'E' followed by '+' or '-' and exactly three digits. Numbers with magnitude less than 10^{**}-999 or larger than $10^{**}999$ are not supported by most computers, but if they are supported then their output is considered to overflow the field and will be output as asterisks.

- On most computers, no more than 15 decimal digits are significant in output, even if more are printed. In any case, output precision cannot be any higher than input precision; few data sets are accurate to 15 digits of precision. Unavoidable loss of precision in intermediate calculations may also reduce precision of output.

- Special values such as infinities and "not a number" values are usually converted to the system-missing value before printing. In a few circumstances, these values are output directly. In fields of width 3 or greater, special values are output as however many characters will fit from `+Infinity` or `-Infinity` for infinities, from `NaN` for "not a number," or from `Unknown` for other values (if any are supported by the system). In fields under 3 columns wide, special values are output as asterisks.

6.7.4.2 Custom Currency Formats

The custom currency formats are closely related to the basic numeric formats, but they allow users to customize the output format. The SET command configures custom currency formats, using the syntax

> SET CC*x*="*string*".

where *x* is A, B, C, D, or E, and *string* is no more than 16 characters long.

string must contain exactly three commas or exactly three periods (but not both), except that a single quote character may be used to "escape" a following comma, period, or single quote. If three commas are used, commas will be used for grouping in output, and a period will be used as the decimal point. Uses of periods reverses these roles.

The commas or periods divide *string* into four fields, called the *negative prefix*, *prefix*, *suffix*, and *negative suffix*, respectively. The prefix and suffix are added to output whenever space is available. The negative prefix and negative suffix are always added to a negative number when the output includes a nonzero digit.

The following syntax shows how custom currency formats could be used to reproduce basic numeric formats:

```
SET CCA="-,,,".  /* Same as COMMA.
SET CCB="-...".  /* Same as DOT.
SET CCC="-,$,,".  /* Same as DOLLAR.
SET CCD="-,,%,".  /* Like PCT, but groups with commas.
```

Here are some more examples of custom currency formats. The final example shows how to use a single quote to escape a delimiter:

```
SET CCA=",EUR,,-".   /* Euro.
SET CCB="(,USD ,,)".  /* US dollar.
SET CCC="-.R$..".    /* Brazilian real.
SET CCD="-,, NIS,".   /* Israel shekel.
SET CCE="-.Rp'. ..".  /* Indonesia Rupiah.
```

These formats would yield the following output:

Format	3145.59	-3145.59
CCA12.2	EUR3,145.59	EUR3,145.59-
CCB14.2	USD 3,145.59	(USD 3,145.59)
CCC11.2	R$3.145,59	-R$3.145,59
CCD13.2	3,145.59 NIS	-3,145.59 NIS
CCE10.0	Rp. 3.146	-Rp. 3.146

The default for all the custom currency formats is '-,,,', equivalent to COMMA format.

6.7.4.3 Legacy Numeric Formats

The N and Z numeric formats provide compatibility with legacy file formats. They have much in common:

- Output is rounded to the nearest representable value, with ties rounded away from zero.

- Numbers too large to display are output as a field filled with asterisks ('*').

- The decimal point is always implicitly the specified number of digits from the right edge of the field, except that Z format input allows an explicit decimal point.

- Scientific notation may not be used.

- The system-missing value is output as a period in a field of spaces. The period is placed just to the right of the implied decimal point in Z format, or at the right end in N format or in Z format if no decimal places are requested. A period is used even if the decimal point character is a comma.

- Field width may range from 1 to 40. Decimal places may range from 0 up to the field width, to a maximum of 16.

- When a legacy numeric format used for input is converted to an output format, it is changed into the equivalent F format. The field width is increased by 1 if any decimal places are specified, to make room for a decimal point. For Z format, the field width is increased by 1 more column, to make room for a negative sign. The output field width is capped at 40 columns.

N Format

The N format supports input and output of fields that contain only digits. On input, leading or trailing spaces, a decimal point, or any other non-digit character causes the field to be read as the system-missing value. As a special exception, an N format used on DATA LIST FREE or DATA LIST LIST is treated as the equivalent F format.

On output, N pads the field on the left with zeros. Negative numbers are output like the system-missing value.

Z Format

The Z format is a "zoned decimal" format used on IBM mainframes. Z format encodes the sign as part of the final digit, which must be one of the following:

```
0123456789
{ABCDEFGHI
}JKLMNOPQR
```

where the characters in each row represent digits 0 through 9 in order. Characters in the first two rows indicate a positive sign; those in the third indicate a negative sign.

On output, Z fields are padded on the left with spaces. On input, leading and trailing spaces are ignored. Any character in an input field other than spaces, the digit characters above, and '.' causes the field to be read as system-missing.

The decimal point character for input and output is always '.', even if the decimal point character is a comma (see [SET DECIMAL], page 159).

Nonzero, negative values output in Z format are marked as negative even when no nonzero digits are output. For example, -0.2 is output in Z1.0 format as 'J'. The "negative zero" value supported by most machines is output as positive.

6.7.4.4 Binary and Hexadecimal Numeric Formats

The binary and hexadecimal formats are primarily designed for compatibility with existing machine formats, not for human readability. All of them therefore have a F format as default output format. Some of these formats are only portable between machines with compatible byte ordering (endianness) or floating-point format.

Binary formats use byte values that in text files are interpreted as special control functions, such as carriage return and line feed. Thus, data in binary formats should not be included in syntax files or read from data files with variable-length records, such as ordinary text files. They may be read from or written to data files with fixed-length records. See Section 8.8 [FILE HANDLE], page 70, for information on working with fixed-length records.

P and PK Formats

These are binary-coded decimal formats, in which every byte (except the last, in P format) represents two decimal digits. The most-significant 4 bits of the first byte is the most-significant decimal digit, the least-significant 4 bits of the first byte is the next decimal digit, and so on.

In P format, the most-significant 4 bits of the last byte are the least-significant decimal digit. The least-significant 4 bits represent the sign: decimal 15 indicates a negative value, decimal 13 indicates a positive value.

Numbers are rounded downward on output. The system-missing value and numbers outside representable range are output as zero.

The maximum field width is 16. Decimal places may range from 0 up to the number of decimal digits represented by the field.

The default output format is an F format with twice the input field width, plus one column for a decimal point (if decimal places were requested).

IB and PIB Formats

These are integer binary formats. IB reads and writes 2's complement binary integers, and PIB reads and writes unsigned binary integers. The byte ordering is by default the host machine's, but SET RIB may be used to select a specific byte ordering for reading (see [SET RIB], page 159) and SET WIB, similarly, for writing (see [SET WIB], page 162).

The maximum field width is 8. Decimal places may range from 0 up to the number of decimal digits in the largest value representable in the field width.

The default output format is an F format whose width is the number of decimal digits in the largest value representable in the field width, plus 1 if the format has decimal places.

RB Format

This is a binary format for real numbers. By default it reads and writes the host machine's floating-point format, but SET RRB may be used to select an alternate floating-point format for reading (see [SET RRB], page 160) and SET WRB, similarly, for writing (see [SET WRB], page 162).

The recommended field width depends on the floating-point format. NATIVE (the default format), IDL, IDB, VD, VG, and ZL formats should use a field width of 8. ISL, ISB, VF, and ZS formats should use a field width of 4. Other field widths will not produce useful results. The maximum field width is 8. No decimal places may be specified.

The default output format is F8.2.

PIBHEX and RBHEX Formats

These are hexadecimal formats, for reading and writing binary formats where each byte has been recoded as a pair of hexadecimal digits.

A hexadecimal field consists solely of hexadecimal digits '0'...'9' and 'A'...'F'. Uppercase and lowercase are accepted on input; output is in uppercase.

Other than the hexadecimal representation, these formats are equivalent to PIB and RB formats, respectively. However, bytes in PIBHEX format are always ordered with the most-significant byte first (big-endian order), regardless of the host machine's native byte order or PSPP settings.

Field widths must be even and between 2 and 16. RBHEX format allows no decimal places; PIBHEX allows as many decimal places as a PIB format with half the given width.

6.7.4.5 Time and Date Formats

In PSPP, a *time* is an interval. The time formats translate between human-friendly descriptions of time intervals and PSPP's internal representation of time intervals, which is simply the number of seconds in the interval. PSPP has two time formats:

Time Format	Template	Example
TIME	`hh:MM:SS.ss`	`04:31:17.01`
DTIME	`DD HH:MM:SS.ss`	`00 04:31:17.01`

A *date* is a moment in the past or the future. Internally, PSPP represents a date as the number of seconds since the *epoch*, midnight, Oct. 14, 1582. The date formats translate between human-readable dates and PSPP's numeric representation of dates and times. PSPP has several date formats:

Date Format	Template	Example
DATE	`dd-mmm-yyyy`	`01-OCT-1978`
ADATE	`mm/dd/yyyy`	`10/01/1978`
EDATE	`dd.mm.yyyy`	`01.10.1978`
JDATE	`yyyyjjj`	`1978274`
SDATE	`yyyy/mm/dd`	`1978/10/01`
QYR	`q Q yyyy`	`3 Q 1978`
MOYR	`mmm yyyy`	`OCT 1978`
WKYR	`ww WK yyyy`	`40 WK 1978`
DATETIME	`dd-mmm-yyyy HH:MM:SS.ss`	`01-OCT-1978 04:31:17.01`

The templates in the preceding tables describe how the time and date formats are input and output:

`dd`   Day of month, from 1 to 31. Always output as two digits.

`mm`
`mmm`   Month. In output, `mm` is output as two digits, `mmm` as the first three letters of an English month name (January, February, ...). In input, both of these formats, plus Roman numerals, are accepted.

`yyyy`   Year. In output, DATETIME always produces a 4-digit year; other formats can produce a 2- or 4-digit year. The century assumed for 2-digit years depends on the EPOCH setting (see [SET EPOCH], page 159). In output, a year outside the epoch causes the whole field to be filled with asterisks ('*').

`jjj`   Day of year (Julian day), from 1 to 366. This is exactly three digits giving the count of days from the start of the year. January 1 is considered day 1.

`q`   Quarter of year, from 1 to 4. Quarters start on January 1, April 1, July 1, and October 1.

`ww`   Week of year, from 1 to 53. Output as exactly two digits. January 1 is the first day of week 1.

`DD`   Count of days, which may be positive or negative. Output as at least two digits.

`hh`   Count of hours, which may be positive or negative. Output as at least two digits.

`HH`   Hour of day, from 0 to 23. Output as exactly two digits.

MM   Minute of hour, from 0 to 59. Output as exactly two digits.

SS.ss   Seconds within minute, from 0 to 59. The integer part is output as exactly two
            digits. On output, seconds and fractional seconds may or may not be included,
            depending on field width and decimal places. On input, seconds and fractional
            seconds are optional. The DECIMAL setting controls the character accepted
            and displayed as the decimal point (see [SET DECIMAL], page 159).

For output, the date and time formats use the delimiters indicated in the table. For
input, date components may be separated by spaces or by one of the characters '–', '/',
'.', or ',', and time components may be separated by spaces, ':', or '.'. On input, the 'Q'
separating quarter from year and the 'WK' separating week from year may be uppercase or
lowercase, and the spaces around them are optional.

On input, all time and date formats accept any amount of leading and trailing white
space.

The maximum width for time and date formats is 40 columns. Minimum input and
output width for each of the time and date formats is shown below:

Format	Min. Input Width	Min. Output Width	Option
DATE	8	9	4-digit year
ADATE	8	8	4-digit year
EDATE	8	8	4-digit year
JDATE	5	5	4-digit year
SDATE	8	8	4-digit year
QYR	4	6	4-digit year
MOYR	6	6	4-digit year
WKYR	6	8	4-digit year
DATETIME	17	17	seconds
TIME	5	5	seconds
DTIME	8	8	seconds

In the table, "Option" describes what increased output width enables:

4-digit year
            A field 2 columns wider than minimum will include a 4-digit year. (DATETIME
            format always includes a 4-digit year.)

seconds   A field 3 columns wider than minimum will include seconds as well as minutes.
            A field 5 columns wider than minimum, or more, can also include a decimal
            point and fractional seconds (but no more than allowed by the format's decimal
            places).

For the time and date formats, the default output format is the same as the input format,
except that PSPP increases the field width, if necessary, to the minimum allowed for output.

Time or dates narrower than the field width are right-justified within the field.

When a time or date exceeds the field width, characters are trimmed from the end until
it fits. This can occur in an unusual situation, e.g. with a year greater than 9999 (which
adds an extra digit), or for a negative value on TIME or DTIME (which adds a leading
minus sign).

The system-missing value is output as a period at the right end of the field.

6.7.4.6 Date Component Formats

The WKDAY and MONTH formats provide input and output for the names of weekdays and months, respectively.

On output, these formats convert a number between 1 and 7, for WKDAY, or between 1 and 12, for MONTH, into the English name of a day or month, respectively. If the name is longer than the field, it is trimmed to fit. If the name is shorter than the field, it is padded on the right with spaces. Values outside the valid range, and the system-missing value, are output as all spaces.

On input, English weekday or month names (in uppercase or lowercase) are converted back to their corresponding numbers. Weekday and month names may be abbreviated to their first 2 or 3 letters, respectively.

The field width may range from 2 to 40, for WKDAY, or from 3 to 40, for MONTH. No decimal places are allowed.

The default output format is the same as the input format.

6.7.4.7 String Formats

The A and AHEX formats are the only ones that may be assigned to string variables. Neither format allows any decimal places.

In A format, the entire field is treated as a string value. The field width may range from 1 to 32,767, the maximum string width. The default output format is the same as the input format.

In AHEX format, the field is composed of characters in a string encoded as hex digit pairs. On output, hex digits are output in uppercase; on input, uppercase and lowercase are both accepted. The default output format is A format with half the input width.

6.7.5 Scratch Variables

Most of the time, variables don't retain their values between cases. Instead, either they're being read from a data file or the active dataset, in which case they assume the value read, or, if created with COMPUTE or another transformation, they're initialized to the system-missing value or to blanks, depending on type.

However, sometimes it's useful to have a variable that keeps its value between cases. You can do this with LEAVE (see Section 11.5 [LEAVE], page 101), or you can use a *scratch variable*. Scratch variables are variables whose names begin with an octothorpe ('#').

Scratch variables have the same properties as variables left with LEAVE: they retain their values between cases, and for the first case they are initialized to 0 or blanks. They have the additional property that they are deleted before the execution of any procedure. For this reason, scratch variables can't be used for analysis. To use a scratch variable in an analysis, use COMPUTE (see Section 12.3 [COMPUTE], page 113) to copy its value into an ordinary variable, then use that ordinary variable in the analysis.

6.8 Files Used by PSPP

PSPP makes use of many files each time it runs. Some of these it reads, some it writes, some it creates. Here is a table listing the most important of these files:

command file

syntax file These names (synonyms) refer to the file that contains instructions that tell PSPP what to do. The syntax file's name is specified on the PSPP command line. Syntax files can also be read with INCLUDE (see Section 16.15 [INCLUDE], page 155).

data file Data files contain raw data in text or binary format. Data can also be embedded in a syntax file with BEGIN DATA and END DATA.

listing file One or more output files are created by PSPP each time it is run. The output files receive the tables and charts produced by statistical procedures. The output files may be in any number of formats, depending on how PSPP is configured.

system file

System files are binary files that store a dictionary and a set of cases. GET and SAVE read and write system files.

portable file

Portable files are files in a text-based format that store a dictionary and a set of cases. IMPORT and EXPORT read and write portable files.

6.9 File Handles

A *file handle* is a reference to a data file, system file, or portable file. Most often, a file handle is specified as the name of a file as a string, that is, enclosed within '' or "".

A file name string that begins or ends with '|' is treated as the name of a command to pipe data to or from. You can use this feature to read data over the network using a program such as 'curl' (e.g. GET '|curl -s -S http://example.com/mydata.sav'), to read compressed data from a file using a program such as 'zcat' (e.g. GET '|zcat mydata.sav.gz'), and for many other purposes.

PSPP also supports declaring named file handles with the FILE HANDLE command. This command associates an identifier of your choice (the file handle's name) with a file. Later, the file handle name can be substituted for the name of the file. When PSPP syntax accesses a file multiple times, declaring a named file handle simplifies updating the syntax later to use a different file. Use of FILE HANDLE is also required to read data files in binary formats. See Section 8.8 [FILE HANDLE], page 70, for more information.

In some circumstances, PSPP must distinguish whether a file handle refers to a system file or a portable file. When this is necessary to read a file, e.g. as an input file for GET or MATCH FILES, PSPP uses the file's contents to decide. In the context of writing a file, e.g. as an output file for SAVE or AGGREGATE, PSPP decides based on the file's name: if it ends in '.por' (with any capitalization), then PSPP writes a portable file; otherwise, PSPP writes a system file.

INLINE is reserved as a file handle name. It refers to the "data file" embedded into the syntax file between BEGIN DATA and END DATA. See Section 8.1 [BEGIN DATA], page 64, for more information.

The file to which a file handle refers may be reassigned on a later FILE HANDLE command if it is first closed using CLOSE FILE HANDLE. See Section 8.2 [CLOSE FILE HANDLE], page 64, for more information.

6.10 Backus-Naur Form

The syntax of some parts of the PSPP language is presented in this manual using the formalism known as *Backus-Naur Form*, or BNF. The following table describes BNF:

- Words in all-uppercase are PSPP keyword tokens. In BNF, these are often called *terminals*. There are some special terminals, which are written in lowercase for clarity:

  number   A real number.

  integer   An integer number.

  string   A string.

  var-name A single variable name.

  =, /, +, -, etc.
  > Operators and punctuators.

  .
  > The end of the command. This is not necessarily an actual dot in the syntax file: See Section 6.2 [Commands], page 29, for more details.

- Other words in all lowercase refer to BNF definitions, called *productions*. These productions are also known as *nonterminals*. Some nonterminals are very common, so they are defined here in English for clarity:

  var-list A list of one or more variable names or the keyword ALL.

  expression
  > An expression. See Chapter 7 [Expressions], page 46, for details.

- '::=' means "is defined as". The left side of '::=' gives the name of the nonterminal being defined. The right side of '::=' gives the definition of that nonterminal. If the right side is empty, then one possible expansion of that nonterminal is nothing. A BNF definition is called a *production*.

- So, the key difference between a terminal and a nonterminal is that a terminal cannot be broken into smaller parts—in fact, every terminal is a single token (see Section 6.1 [Tokens], page 28). On the other hand, nonterminals are composed of a (possibly empty) sequence of terminals and nonterminals. Thus, terminals indicate the deepest level of syntax description. (In parsing theory, terminals are the leaves of the parse tree; nonterminals form the branches.)

- The first nonterminal defined in a set of productions is called the *start symbol*. The start symbol defines the entire syntax for that command.

7 Mathematical Expressions

Expressions share a common syntax each place they appear in PSPP commands. Expressions are made up of *operands*, which can be numbers, strings, or variable names, separated by *operators*. There are five types of operators: grouping, arithmetic, logical, relational, and functions.

Every operator takes one or more operands as input and yields exactly one result as output. Depending on the operator, operands accept strings or numbers as operands. With few exceptions, operands may be full-fledged expressions in themselves.

7.1 Boolean Values

Some PSPP operators and expressions work with Boolean values, which represent true/false conditions. Booleans have only three possible values: 0 (false), 1 (true), and system-missing (unknown). System-missing is neither true nor false and indicates that the true value is unknown.

Boolean-typed operands or function arguments must take on one of these three values. Other values are considered false, but provoke a warning when the expression is evaluated.

Strings and Booleans are not compatible, and neither may be used in place of the other.

7.2 Missing Values in Expressions

Most numeric operators yield system-missing when given any system-missing operand. A string operator given any system-missing operand typically results in the empty string. Exceptions are listed under particular operator descriptions.

String user-missing values are not treated specially in expressions.

User-missing values for numeric variables are always transformed into the system-missing value, except inside the arguments to the VALUE and SYSMIS functions.

The missing-value functions can be used to precisely control how missing values are treated in expressions. See Section 7.7.4 [Missing Value Functions], page 49, for more details.

7.3 Grouping Operators

Parentheses ('()') are the grouping operators. Surround an expression with parentheses to force early evaluation.

Parentheses also surround the arguments to functions, but in that situation they act as punctuators, not as operators.

7.4 Arithmetic Operators

The arithmetic operators take numeric operands and produce numeric results.

a + b   Yields the sum of a and b.

a - b   Subtracts b from a and yields the difference.

a * b   Yields the product of a and b. If either a or b is 0, then the result is 0, even if the other operand is missing.

a / b   Divides a by b and yields the quotient. If a is 0, then the result is 0, even if b is missing. If b is zero, the result is system-missing.

a ** b   Yields the result of raising a to the power b. If a is negative and b is not an integer, the result is system-missing. The result of 0**0 is system-missing as well.

- a   Reverses the sign of a.

7.5 Logical Operators

The logical operators take logical operands and produce logical results, meaning "true or false." Logical operators are not true Boolean operators because they may also result in a system-missing value. See Section 7.1 [Boolean Values], page 46, for more information.

a AND b

a & b   True if both a and b are true, false otherwise. If one operand is false, the result is false even if the other is missing. If both operands are missing, the result is missing.

a OR b

a | b   True if at least one of a and b is true. If one operand is true, the result is true even if the other operand is missing. If both operands are missing, the result is missing.

NOT a

~ a   True if a is false. If the operand is missing, then the result is missing.

7.6 Relational Operators

The relational operators take numeric or string operands and produce Boolean results.

Strings cannot be compared to numbers. When strings of different lengths are compared, the shorter string is right-padded with spaces to match the length of the longer string.

The results of string comparisons, other than tests for equality or inequality, depend on the character set in use. String comparisons are case-sensitive.

a EQ b

a = b   True if a is equal to b.

a LE b

a <= b   True if a is less than or equal to b.

a LT b

a < b   True if a is less than b.

a GE b

a >= b   True if a is greater than or equal to b.

a GT b

a > b   True if a is greater than b.

a NE b

a ~= b

a <> b   True if a is not equal to b.

7.7 Functions

PSPP functions provide mathematical abilities above and beyond those possible using simple operators. Functions have a common syntax: each is composed of a function name followed by a left parenthesis, one or more arguments, and a right parenthesis.

Function names are not reserved. Their names are specially treated only when followed by a left parenthesis, so that 'EXP(10)' refers to the constant value *e* raised to the 10th power, but 'EXP' by itself refers to the value of a variable called EXP.

The sections below describe each function in detail.

7.7.1 Mathematical Functions

Advanced mathematical functions take numeric arguments and produce numeric results.

EXP (*exponent*)   [Function]
> Returns *e* (approximately 2.71828) raised to power *exponent*.

LG10 (*number*)   [Function]
> Takes the base-10 logarithm of *number*. If *number* is not positive, the result is system-missing.

LN (*number*)   [Function]
> Takes the base-*e* logarithm of *number*. If *number* is not positive, the result is system-missing.

LNGAMMA (*number*)   [Function]
> Yields the base-*e* logarithm of the complete gamma of *number*. If *number* is a negative integer, the result is system-missing.

SQRT (*number*)   [Function]
> Takes the square root of *number*. If *number* is negative, the result is system-missing.

7.7.2 Miscellaneous Mathematical Functions

Miscellaneous mathematical functions take numeric arguments and produce numeric results.

ABS (*number*)   [Function]
> Results in the absolute value of *number*.

MOD (*numerator*, *denominator*)   [Function]
> Returns the remainder (modulus) of *numerator* divided by *denominator*. If *numerator* is 0, then the result is 0, even if *denominator* is missing. If *denominator* is 0, the result is system-missing.

MOD10 (*number*)   [Function]
> Returns the remainder when *number* is divided by 10. If *number* is negative, MOD10(*number*) is negative or zero.

RND (*number* [, *mult*[, *fuzzbits*]])   [Function]
> Rounds *number* and rounds it to a multiple of *mult* (by default 1). Halves are rounded away from zero, as are values that fall short of halves by less than *fuzzbits* of errors in the least-significant bits of *number*. If *fuzzbits* is not specified then the default is taken from SET FUZZBITS (see [SET FUZZBITS], page 161), which is 6 unless overridden.

TRUNC (*number*)   [Function]
>   Discards the fractional part of *number*; that is, rounds *number* towards zero.

7.7.3 Trigonometric Functions

Trigonometric functions take numeric arguments and produce numeric results.

ARCOS (*number*)   [Function]
ACOS (*number*)   [Function]
>   Takes the arccosine, in radians, of *number*. Results in system-missing if *number* is
>   not between -1 and 1 inclusive. This function is a PSPP extension.

ARSIN (*number*)   [Function]
ASIN (*number*)   [Function]
>   Takes the arcsine, in radians, of *number*. Results in system-missing if *number* is not
>   between -1 and 1 inclusive.

ARTAN (*number*)   [Function]
ATAN (*number*)   [Function]
>   Takes the arctangent, in radians, of *number*.

COS (*angle*)   [Function]
>   Takes the cosine of *angle* which should be in radians.

SIN (*angle*)   [Function]
>   Takes the sine of *angle* which should be in radians.

TAN (*angle*)   [Function]
>   Takes the tangent of *angle* which should be in radians. Results in system-missing at
>   values of *angle* that are too close to odd multiples of $\pi/2$. Portability: none.

7.7.4 Missing-Value Functions

Missing-value functions take various numeric arguments and yield various types of results.
Except where otherwise stated below, the normal rules of evaluation apply within expression
arguments to these functions. In particular, user-missing values for numeric variables are
converted to system-missing values.

MISSING (*expr*)   [Function]
>   Returns 1 if *expr* has the system-missing value, 0 otherwise.

NMISS (*expr* [, *expr*]...)   [Function]
>   Each argument must be a numeric expression. Returns the number of system-missing
>   values in the list, which may include variable ranges using the *var1* TO *var2* syntax.

NVALID (*expr* [, *expr*]...)   [Function]
>   Each argument must be a numeric expression. Returns the number of values in the
>   list that are not system-missing. The list may include variable ranges using the *var1*
>   TO *var2* syntax.

SYSMIS (*expr*)   [Function]
>   When *expr* is simply the name of a numeric variable, returns 1 if the variable has
>   the system-missing value, 0 if it is user-missing or not missing. If given *expr* takes
>   another form, results in 1 if the value is system-missing, 0 otherwise.

VALUE (`variable`)   [Function]

> Prevents the user-missing values of *variable* from being transformed into system-missing values, and always results in the actual value of *variable*, whether it is valid, user-missing, or system-missing.

7.7.5 Set-Membership Functions

Set membership functions determine whether a value is a member of a set. They take a set of numeric arguments or a set of string arguments, and produce Boolean results.

String comparisons are performed according to the rules given in Section 7.6 [Relational Operators], page 47.

ANY (`value`, `set` [, `set`]...)   [Function]

> Results in true if *value* is equal to any of the *set* values. Otherwise, results in false. If *value* is system-missing, returns system-missing. System-missing values in *set* do not cause /NAME/ to return system-missing.

RANGE (`value`, `low`, `high` [, `low`, `high`]...)   [Function]

> Results in true if *value* is in any of the intervals bounded by *low* and *high* inclusive. Otherwise, results in false. Each *low* must be less than or equal to its corresponding *high* value. *low* and *high* must be given in pairs. If *value* is system-missing, returns system-missing. System-missing values in *set* do not cause /NAME/ to return system-missing.

7.7.6 Statistical Functions

Statistical functions compute descriptive statistics on a list of values. Some statistics can be computed on numeric or string values; other can only be computed on numeric values. Their results have the same type as their arguments. The current case's weighting factor (see Section 13.7 [WEIGHT], page 123) has no effect on statistical functions.

These functions' argument lists may include entire ranges of variables using the `var1 TO var2` syntax.

Unlike most functions, statistical functions can return non-missing values even when some of their arguments are missing. Most statistical functions, by default, require only 1 non-missing value to have a non-missing return, but /NAME/, /NAME/, and /NAME/ require 2. These defaults can be increased (but not decreased) by appending a dot and the minimum number of valid arguments to the function name. For example, `MEAN.3(X, Y, Z)` would only return non-missing if all of 'X', 'Y', and 'Z' were valid.

CFVAR (`number`, `number`[, ...])   [Function]

> Results in the coefficient of variation of the values of *number*. (The coefficient of variation is the standard deviation divided by the mean.)

MAX (`value`, `value`[, ...])   [Function]

> Results in the value of the greatest *value*. The *values* may be numeric or string.

MEAN (`number`, `number`[, ...])   [Function]

> Results in the mean of the values of *number*.

MIN (`number`, `number`[, ...])   [Function]

> Results in the value of the least *value*. The *values* may be numeric or string.

SD (*number*, *number*[, ...])   [Function]
> Results in the standard deviation of the values of *number*.

SUM (*number*, *number*[, ...])   [Function]
> Results in the sum of the values of *number*.

VARIANCE (*number*, *number*[, ...])   [Function]
> Results in the variance of the values of *number*.

7.7.7 String Functions

String functions take various arguments and return various results.

CONCAT (*string*, *string*[, ...])   [Function]
> Returns a string consisting of each *string* in sequence. CONCAT("abc", "def", "ghi") has a value of "abcdefghi". The resultant string is truncated to a maximum of 255 characters.

INDEX (*haystack*, *needle*)   [Function]
> Returns a positive integer indicating the position of the first occurrence of *needle* in *haystack*. Returns 0 if *haystack* does not contain *needle*. Returns system-missing if *needle* is an empty string.

INDEX (*haystack*, *needles*, *needle_len*)   [Function]
> Divides *needles* into one or more needles, each with length *needle_len*. Searches *haystack* for the first occurrence of each needle, and returns the smallest value. Returns 0 if *haystack* does not contain any part in *needle*. It is an error if *needle_len* does not evenly divide the length of *needles*. Returns system-missing if *needles* is an empty string.

LENGTH (*string*)   [Function]
> Returns the number of characters in *string*.

LOWER (*string*)   [Function]
> Returns a string identical to *string* except that all uppercase letters are changed to lowercase letters. The definitions of "uppercase" and "lowercase" are system-dependent.

LPAD (*string*, *length*)   [Function]
> If *string* is at least *length* characters in length, returns *string* unchanged. Otherwise, returns *string* padded with spaces on the left side to length *length*. Returns an empty string if *length* is system-missing, negative, or greater than 255.

LPAD (*string*, *length*, *padding*)   [Function]
> If *string* is at least *length* characters in length, returns *string* unchanged. Otherwise, returns *string* padded with *padding* on the left side to length *length*. Returns an empty string if *length* is system-missing, negative, or greater than 255, or if *padding* does not contain exactly one character.

LTRIM (*string*)   [Function]
> Returns *string*, after removing leading spaces. Other white space, such as tabs, carriage returns, line feeds, and vertical tabs, is not removed.

LTRIM (*string*, *padding*) [Function]

Returns *string*, after removing leading *padding* characters. If *padding* does not contain exactly one character, returns an empty string.

NUMBER (*string*, *format*) [Function]

Returns the number produced when *string* is interpreted according to format specifier *format*. If the format width w is less than the length of *string*, then only the first w characters in *string* are used, e.g. NUMBER("123", F3.0) and NUMBER("1234", F3.0) both have value 123. If w is greater than *string*'s length, then it is treated as if it were right-padded with spaces. If *string* is not in the correct format for *format*, system-missing is returned.

RINDEX (*string*, *format*) [Function]

Returns a positive integer indicating the position of the last occurrence of *needle* in *haystack*. Returns 0 if *haystack* does not contain *needle*. Returns system-missing if *needle* is an empty string.

RINDEX (*haystack*, *needle*, *needle_len*) [Function]

Divides *needle* into parts, each with length *needle_len*. Searches *haystack* for the last occurrence of each part, and returns the largest value. Returns 0 if *haystack* does not contain any part in *needle*. It is an error if *needle_len* does not evenly divide the length of *needle*. Returns system-missing if *needle* is an empty string.

RPAD (*string*, *length*) [Function]

If *string* is at least *length* characters in length, returns *string* unchanged. Otherwise, returns *string* padded with spaces on the right to length *length*. Returns an empty string if *length* is system-missing, negative, or greater than 255.

RPAD (*string*, *length*, *padding*) [Function]

If *string* is at least *length* characters in length, returns *string* unchanged. Otherwise, returns *string* padded with *padding* on the right to length *length*. Returns an empty string if *length* is system-missing, negative, or greater than 255, or if *padding* does not contain exactly one character.

RTRIM (*string*) [Function]

Returns *string*, after removing trailing spaces. Other types of white space are not removed.

RTRIM (*string*, *padding*) [Function]

Returns *string*, after removing trailing *padding* characters. If *padding* does not contain exactly one character, returns an empty string.

STRING (*number*, *format*) [Function]

Returns a string corresponding to *number* in the format given by format specifier *format*. For example, STRING(123.56, F5.1) has the value "123.6".

SUBSTR (*string*, *start*) [Function]

Returns a string consisting of the value of *string* from position *start* onward. Returns an empty string if *start* is system-missing, less than 1, or greater than the length of *string*.

SUBSTR (*string*, *start*, *count*)   [Function]
> Returns a string consisting of the first *count* characters from *string* beginning at position *start*. Returns an empty string if *start* or *count* is system-missing, if *start* is less than 1 or greater than the number of characters in *string*, or if *count* is less than 1. Returns a string shorter than *count* characters if *start* + *count* - 1 is greater than the number of characters in *string*. Examples: SUBSTR("abcdefg", 3, 2) has value "cd"; SUBSTR("nonsense", 4, 10) has the value "sense".

UPCASE (*string*)   [Function]
> Returns *string*, changing lowercase letters to uppercase letters.

7.7.8 Time & Date Functions

For compatibility, PSPP considers dates before 15 Oct 1582 invalid. Most time and date functions will not accept earlier dates.

7.7.8.1 How times & dates are defined and represented

Times and dates are handled by PSPP as single numbers. A *time* is an interval. PSPP measures times in seconds. Thus, the following intervals correspond with the numeric values given:

10 minutes	600
1 hour	3,600
1 day, 3 hours, 10 seconds	97,210
40 days	3,456,000

A *date*, on the other hand, is a particular instant in the past or the future. PSPP represents a date as a number of seconds since midnight preceding 14 Oct 1582. Because midnight preceding the dates given below correspond with the numeric PSPP dates given:

15 Oct 1582	86,400
4 Jul 1776	6,113,318,400
1 Jan 1900	10,010,390,400
1 Oct 1978	12,495,427,200
24 Aug 1995	13,028,601,600

7.7.8.2 Functions that Produce Times

These functions take numeric arguments and return numeric values that represent times.

TIME.DAYS (*ndays*)   [Function]
> Returns a time corresponding to *ndays* days.

TIME.HMS (*nhours*, *nmins*, *nsecs*)   [Function]
> Returns a time corresponding to *nhours* hours, *nmins* minutes, and *nsecs* seconds. The arguments may not have mixed signs: if any of them are positive, then none may be negative, and vice versa.

7.7.8.3 Functions that Examine Times

These functions take numeric arguments in PSPP time format and give numeric results.

CTIME.DAYS (*time*)   [Function]
> Results in the number of days and fractional days in *time*.

CTIME.HOURS (*time*)   [Function]
>   Results in the number of hours and fractional hours in *time*.

CTIME.MINUTES (*time*)   [Function]
>   Results in the number of minutes and fractional minutes in *time*.

CTIME.SECONDS (*time*)   [Function]
>   Results in the number of seconds and fractional seconds in *time*. (CTIME.SECONDS
>   does nothing; CTIME.SECONDS(*x*) is equivalent to *x*.)

7.7.8.4 Functions that Produce Dates

These functions take numeric arguments and give numeric results that represent dates.
Arguments taken by these functions are:

day   Refers to a day of the month between 1 and 31. Day 0 is also accepted and
             refers to the final day of the previous month. Days 29, 30, and 31 are accepted
             even in months that have fewer days and refer to a day near the beginning of
             the following month.

month   Refers to a month of the year between 1 and 12. Months 0 and 13 are also
             accepted and refer to the last month of the preceding year and the first month
             of the following year, respectively.

quarter   Refers to a quarter of the year between 1 and 4. The quarters of the year begin
             on the first day of months 1, 4, 7, and 10.

week   Refers to a week of the year between 1 and 53.

yday   Refers to a day of the year between 1 and 366.

year   Refers to a year, 1582 or greater. Years between 0 and 99 are treated according
             to the epoch set on SET EPOCH, by default beginning 69 years before the
             current date (see [SET EPOCH], page 159).

If these functions' arguments are out-of-range, they are correctly normalized before conversion to date format. Non-integers are rounded toward zero.

DATE.DMY (*day*, *month*, *year*)   [Function]
DATE.MDY (*month*, *day*, *year*)   [Function]
>   Results in a date value corresponding to the midnight before day *day* of month *month*
>   of year *year*.

DATE.MOYR (*month*, *year*)   [Function]
>   Results in a date value corresponding to the midnight before the first day of month
>   *month* of year *year*.

DATE.QYR (*quarter*, *year*)   [Function]
>   Results in a date value corresponding to the midnight before the first day of quarter
>   *quarter* of year *year*.

DATE.WKYR (*week*, *year*)   [Function]
>   Results in a date value corresponding to the midnight before the first day of week
>   *week* of year *year*.

DATE.YRDAY (*year*, *yday*)   [Function]
> Results in a date value corresponding to the day *yday* of year *year*.

7.7.8.5 Functions that Examine Dates

These functions take numeric arguments in PSPP date or time format and give numeric results. These names are used for arguments:

date   A numeric value in PSPP date format.

time   A numeric value in PSPP time format.

time-or-date
> A numeric value in PSPP time or date format.

XDATE.DATE (*time-or-date*)   [Function]
> For a time, results in the time corresponding to the number of whole days *date-or-time* includes. For a date, results in the date corresponding to the latest midnight at or before *date-or-time*; that is, gives the date that *date-or-time* is in.

XDATE.HOUR (*time-or-date*)   [Function]
> For a time, results in the number of whole hours beyond the number of whole days represented by *date-or-time*. For a date, results in the hour (as an integer between 0 and 23) corresponding to *date-or-time*.

XDATE.JDAY (*date*)   [Function]
> Results in the day of the year (as an integer between 1 and 366) corresponding to *date*.

XDATE.MDAY (*date*)   [Function]
> Results in the day of the month (as an integer between 1 and 31) corresponding to *date*.

XDATE.MINUTE (*time-or-date*)   [Function]
> Results in the number of minutes (as an integer between 0 and 59) after the last hour in *time-or-date*.

XDATE.MONTH (*date*)   [Function]
> Results in the month of the year (as an integer between 1 and 12) corresponding to *date*.

XDATE.QUARTER (*date*)   [Function]
> Results in the quarter of the year (as an integer between 1 and 4) corresponding to *date*.

XDATE.SECOND (*time-or-date*)   [Function]
> Results in the number of whole seconds after the last whole minute (as an integer between 0 and 59) in *time-or-date*.

XDATE.TDAY (*date*)   [Function]
> Results in the number of whole days from 14 Oct 1582 to *date*.

XDATE.TIME (*date*)   [Function]

> Results in the time of day at the instant corresponding to *date*, as a time value. This is the number of seconds since midnight on the day corresponding to *date*.

XDATE.WEEK (*date*)   [Function]

> Results in the week of the year (as an integer between 1 and 53) corresponding to *date*.

XDATE.WKDAY (*date*)   [Function]

> Results in the day of week (as an integer between 1 and 7) corresponding to *date*, where 1 represents Sunday.

XDATE.YEAR (*date*)   [Function]

> Returns the year (as an integer 1582 or greater) corresponding to *date*.

7.7.8.6 Time and Date Arithmetic

Ordinary arithmetic operations on dates and times often produce sensible results. Adding a time to, or subtracting one from, a date produces a new date that much earlier or later. The difference of two dates yields the time between those dates. Adding two times produces the combined time. Multiplying a time by a scalar produces a time that many times longer. Since times and dates are just numbers, the ordinary addition and subtraction operators are employed for these purposes.

Adding two dates does not produce a useful result.

Dates and times may have very large values. Thus, it is not a good idea to take powers of these values; also, the accuracy of some procedures may be affected. If necessary, convert times or dates in seconds to some other unit, like days or years, before performing analysis.

PSPP supplies a few functions for date arithmetic:

DATEDIFF (*date2*, *date1*, *unit*)   [Function]

> Returns the span of time from *date1* to *date2* in terms of *unit*, which must be a quoted string, one of 'years', 'quarters', 'months', 'weeks', 'days', 'hours', 'minutes', and 'seconds'. The result is an integer, truncated toward zero.

> One year is considered to span from a given date to the same month, day, and time of day the next year. Thus, from Jan. 1 of one year to Jan. 1 the next year is considered to be a full year, but Feb. 29 of a leap year to the following Feb. 28 is not. Similarly, one month spans from a given day of the month to the same day of the following month. Thus, there is never a full month from Jan. 31 of a given year to any day in the following February.

DATESUM (*date*, *quantity*, *unit*[, *method*])   [Function]

> Returns *date* advanced by the given *quantity* of the specified *unit*, which must be one of the strings 'years', 'quarters', 'months', 'weeks', 'days', 'hours', 'minutes', and 'seconds'.

> When *unit* is 'years', 'quarters', or 'months', only the integer part of *quantity* is considered. Adding one of these units can cause the day of the month to exceed the number of days in the month. In this case, the *method* comes into play: if it is omitted or specified as 'closest' (as a quoted string), then the resulting day is the

last day of the month; otherwise, if it is specified as 'rollover', then the extra days roll over into the following month.

When *unit* is 'weeks', 'days', 'hours', 'minutes', or 'seconds', the *quantity* is not rounded to an integer and *method*, if specified, is ignored.

7.7.9 Miscellaneous Functions

LAG (`variable`[, `n`])   [Function]

variable must be a numeric or string variable name. LAG yields the value of that variable for the case *n* before the current one. Results in system-missing (for numeric variables) or blanks (for string variables) for the first *n* cases.

LAG obtains values from the cases that become the new active dataset after a procedure executes. Thus, LAG will not return values from cases dropped by transformations such as SELECT IF, and transformations like COMPUTE that modify data will change the values returned by LAG. These are both the case whether these transformations precede or follow the use of LAG.

If LAG is used before TEMPORARY, then the values it returns are those in cases just before TEMPORARY. LAG may not be used after TEMPORARY.

If omitted, *ncases* defaults to 1. Otherwise, *ncases* must be a small positive constant integer. There is no explicit limit, but use of a large value will increase memory consumption.

YRMODA (`year`, `month`, `day`)   [Function]

year is a year, either between 0 and 99 or at least 1582. Unlike other PSPP date functions, years between 0 and 99 always correspond to 1900 through 1999. *month* is a month between 1 and 13. *day* is a day between 0 and 31. A *day* of 0 refers to the last day of the previous month, and a *month* of 13 refers to the first month of the next year. *year* must be in range. *year*, *month*, and *day* must all be integers.

YRMODA results in the number of days between 15 Oct 1582 and the date specified, plus one. The date passed to YRMODA must be on or after 15 Oct 1582. 15 Oct 1582 has a value of 1.

VALUELABEL (`variable`)   [Function]

Returns a string matching the label associated with the current value of *variable*. If the current value of *variable* has no associated label, then this function returns the empty string. *variable* may be a numeric or string variable.

7.7.10 Statistical Distribution Functions

PSPP can calculate several functions of standard statistical distributions. These functions are named systematically based on the function and the distribution. The table below describes the statistical distribution functions in general:

PDF.*dist* (x[, *param*...])

Probability density function for *dist*. The domain of x depends on *dist*. For continuous distributions, the result is the density of the probability function at x, and the range is nonnegative real numbers. For discrete distributions, the result is the probability of x.

CDF.*dist* (x[, *param*. . .])
>   Cumulative distribution function for *dist*, that is, the probability that a random variate drawn from the distribution is less than x. The domain of x depends *dist*. The result is a probability.

SIG.*dist* (x[, *param*. . .)
>   Tail probability function for *dist*, that is, the probability that a random variate drawn from the distribution is greater than x. The domain of x depends *dist*. The result is a probability. Only a few distributions include an /NAME/ function.

IDF.*dist* (p[, *param*. . .])
>   Inverse distribution function for *dist*, the value of x for which the CDF would yield p. The value of p is a probability. The range depends on *dist* and is identical to the domain for the corresponding CDF.

RV.*dist* ([*param*. . .])
>   Random variate function for *dist*. The range depends on the distribution.

NPDF.*dist* (x[, *param*. . .])
>   Noncentral probability density function. The result is the density of the given noncentral distribution at x. The domain of x depends on *dist*. The range is nonnegative real numbers. Only a few distributions include an /NAME/ function.

NCDF.*dist* (x[, *param*. . .])
>   Noncentral cumulative distribution function for *dist*, that is, the probability that a random variate drawn from the given noncentral distribution is less than x. The domain of x depends *dist*. The result is a probability. Only a few distributions include an NCDF function.

The individual distributions are described individually below.

7.7.10.1 Continuous Distributions

The following continuous distributions are available:

PDF.BETA (x)   [Function]
CDF.BETA (x, a, b)   [Function]
IDF.BETA (p, a, b)   [Function]
RV.BETA (a, b)   [Function]
NPDF.BETA (x, a, b, lambda)   [Function]
NCDF.BETA (x, a, b, lambda)   [Function]
>   Beta distribution with shape parameters a and b. The noncentral distribution takes an additional parameter *lambda*. Constraints: a > 0, b > 0, *lambda* >= 0, 0 <= x <= 1, 0 <= p <= 1.

PDF.BVNOR (x0, x1, rho)   [Function]
CDF.VBNOR (x0, x1, rho)   [Function]
>   Bivariate normal distribution of two standard normal variables with correlation coefficient *rho*. Two variates x0 and x1 must be provided. Constraints: 0 <= *rho* <= 1, 0 <= p <= 1.

PDF.CAUCHY (x, a, b)   [Function]
CDF.CAUCHY (x, a, b)   [Function]
IDF.CAUCHY (p, a, b)   [Function]
RV.CAUCHY (a, b)   [Function]
    Cauchy distribution with location parameter a and scale parameter b. Constraints: $b > 0$, $0 < p < 1$.

CDF.CHISQ (x, df)   [Function]
SIG.CHISQ (x, df)   [Function]
IDF.CHISQ (p, df)   [Function]
RV.CHISQ (df)   [Function]
NCDF.CHISQ (x, df, $lambda$)   [Function]
    Chi-squared distribution with df degrees of freedom. The noncentral distribution takes an additional parameter $lambda$. Constraints: $df > 0$, $lambda > 0$, $x >= 0$, $0 <= p < 1$.

PDF.EXP (x, a)   [Function]
CDF.EXP (x, a)   [Function]
IDF.EXP (p, a)   [Function]
RV.EXP (a)   [Function]
    Exponential distribution with scale parameter a. The inverse of a represents the rate of decay. Constraints: $a > 0$, $x >= 0$, $0 <= p < 1$.

PDF.XPOWER (x, a, b)   [Function]
RV.XPOWER (a, b)   [Function]
    Exponential power distribution with positive scale parameter a and nonnegative power parameter b. Constraints: $a > 0$, $b >= 0$, $x >= 0$, $0 <= p <= 1$. This distribution is a PSPP extension.

PDF.F (x, $df1$, $df2$)   [Function]
CDF.F (x, $df1$, $df2$)   [Function]
SIG.F (x, $df1$, $df2$)   [Function]
IDF.F (p, $df1$, $df2$)   [Function]
RV.F ($df1$, $df2$)   [Function]
    F-distribution of two chi-squared deviates with $df1$ and $df2$ degrees of freedom. The noncentral distribution takes an additional parameter $lambda$. Constraints: $df1 > 0$, $df2 > 0$, $lambda >= 0$, $x >= 0$, $0 <= p < 1$.

PDF.GAMMA (x, a, b)   [Function]
CDF.GAMMA (x, a, b)   [Function]
IDF.GAMMA (p, a, b)   [Function]
RV.GAMMA (a, b)   [Function]
    Gamma distribution with shape parameter a and scale parameter b. Constraints: $a > 0$, $b > 0$, $x >= 0$, $0 <= p < 1$.

PDF.LANDAU (x)   [Function]
RV.LANDAU ()   [Function]
    Landau distribution.

PDF.LAPLACE (*x*, *a*, *b*)   [Function]
CDF.LAPLACE (*x*, *a*, *b*)   [Function]
IDF.LAPLACE (*p*, *a*, *b*)   [Function]
RV.LAPLACE (*a*, *b*)   [Function]
> Laplace distribution with location parameter *a* and scale parameter *b*. Constraints: $b > 0, 0 < p < 1$.

RV.LEVY (*c*, *alpha*)   [Function]
> Levy symmetric alpha-stable distribution with scale *c* and exponent *alpha*. Constraints: $0 < alpha <= 2$.

RV.LVSKEW (*c*, *alpha*, *beta*)   [Function]
> Levy skew alpha-stable distribution with scale *c*, exponent *alpha*, and skewness parameter *beta*. Constraints: $0 < alpha <= 2, -1 <= beta <= 1$.

PDF.LOGISTIC (*x*, *a*, *b*)   [Function]
CDF.LOGISTIC (*x*, *a*, *b*)   [Function]
IDF.LOGISTIC (*p*, *a*, *b*)   [Function]
RV.LOGISTIC (*a*, *b*)   [Function]
> Logistic distribution with location parameter *a* and scale parameter *b*. Constraints: $b > 0, 0 < p < 1$.

PDF.LNORMAL (*x*, *a*, *b*)   [Function]
CDF.LNORMAL (*x*, *a*, *b*)   [Function]
IDF.LNORMAL (*p*, *a*, *b*)   [Function]
RV.LNORMAL (*a*, *b*)   [Function]
> Lognormal distribution with parameters *a* and *b*. Constraints: $a > 0, b > 0, x >= 0, 0 <= p < 1$.

PDF.NORMAL (*x*, *mu*, *sigma*)   [Function]
CDF.NORMAL (*x*, *mu*, *sigma*)   [Function]
IDF.NORMAL (*p*, *mu*, *sigma*)   [Function]
RV.NORMAL (*mu*, *sigma*)   [Function]
> Normal distribution with mean *mu* and standard deviation *sigma*. Constraints: $b > 0, 0 < p < 1$. Three additional functions are available as shorthand:

> CDFNORM (*x*)   [Function]
>> Equivalent to CDF.NORMAL(*x*, 0, 1).

> PROBIT (*p*)   [Function]
>> Equivalent to IDF.NORMAL(*p*, 0, 1).

> NORMAL (*sigma*)   [Function]
>> Equivalent to RV.NORMAL(0, *sigma*).

PDF.NTAIL (*x*, *a*, *sigma*)   [Function]
RV.NTAIL (*a*, *sigma*)   [Function]
> Normal tail distribution with lower limit *a* and standard deviation *sigma*. This distribution is a PSPP extension. Constraints: $a > 0, x > a, 0 < p < 1$.

PDF.PARETO (*x*, a, b)   [Function]
CDF.PARETO (*x*, a, b)   [Function]
IDF.PARETO (*p*, a, b)   [Function]
RV.PARETO (a, b)   [Function]

> Pareto distribution with threshold parameter *a* and shape parameter *b*. Constraints: a > 0, b > 0, x >= a, 0 <= p < 1.

PDF.RAYLEIGH (*x*, *sigma*)   [Function]
CDF.RAYLEIGH (*x*, *sigma*)   [Function]
IDF.RAYLEIGH (*p*, *sigma*)   [Function]
RV.RAYLEIGH (*sigma*)   [Function]

> Rayleigh distribution with scale parameter *sigma*. This distribution is a PSPP extension. Constraints: *sigma* > 0, x > 0.

PDF.RTAIL (*x*, a, *sigma*)   [Function]
RV.RTAIL (a, *sigma*)   [Function]

> Rayleigh tail distribution with lower limit *a* and scale parameter *sigma*. This distribution is a PSPP extension. Constraints: a > 0, *sigma* > 0, x > a.

PDF.T (*x*, *df*)   [Function]
CDF.T (*x*, *df*)   [Function]
IDF.T (*p*, *df*)   [Function]
RV.T (*df*)   [Function]

> T-distribution with *df* degrees of freedom. The noncentral distribution takes an additional parameter *lambda*. Constraints: *df* > 0, 0 < p < 1.

PDF.T1G (*x*, a, b)   [Function]
CDF.T1G (*x*, a, b)   [Function]
IDF.T1G (*p*, a, b)   [Function]

> Type-1 Gumbel distribution with parameters *a* and *b*. This distribution is a PSPP extension. Constraints: 0 < p < 1.

PDF.T2G (*x*, a, b)   [Function]
CDF.T2G (*x*, a, b)   [Function]
IDF.T2G (*p*, a, b)   [Function]

> Type-2 Gumbel distribution with parameters *a* and *b*. This distribution is a PSPP extension. Constraints: x > 0, 0 < p < 1.

PDF.UNIFORM (*x*, a, b)   [Function]
CDF.UNIFORM (*x*, a, b)   [Function]
IDF.UNIFORM (*p*, a, b)   [Function]
RV.UNIFORM (a, b)   [Function]

> Uniform distribution with parameters *a* and *b*. Constraints: a <= x <= b, 0 <= p <= 1. An additional function is available as shorthand:

> UNIFORM (*b*)   [Function]
> > Equivalent to RV.UNIFORM(0, *b*).

PDF.WEIBULL (*x*, a, b)   [Function]
CDF.WEIBULL (*x*, a, b)   [Function]

`IDF.WEIBULL` (*p, a, b*)   [Function]
`RV.WEIBULL` (*a, b*)   [Function]
>   Weibull distribution with parameters *a* and *b*. Constraints: $a > 0$, $b > 0$, $x \geq 0$, $0 \leq p < 1$.

7.7.10.2 Discrete Distributions

The following discrete distributions are available:

`PDF.BERNOULLI` (*x*)   [Function]
`CDF.BERNOULLI` (*x, p*)   [Function]
`RV.BERNOULLI` (*p*)   [Function]
>   Bernoulli distribution with probability of success *p*. Constraints: $x = 0$ or 1, $0 \leq p \leq 1$.

`PDF.BINOM` (*x, n, p*)   [Function]
`CDF.BINOM` (*x, n, p*)   [Function]
`RV.BINOM` (*n, p*)   [Function]
>   Binomial distribution with *n* trials and probability of success *p*. Constraints: integer $n > 0$, $0 \leq p \leq 1$, integer $x \leq n$.

`PDF.GEOM` (*x, n, p*)   [Function]
`CDF.GEOM` (*x, n, p*)   [Function]
`RV.GEOM` (*n, p*)   [Function]
>   Geometric distribution with probability of success *p*. Constraints: $0 \leq p \leq 1$, integer $x > 0$.

`PDF.HYPER` (*x, a, b, c*)   [Function]
`CDF.HYPER` (*x, a, b, c*)   [Function]
`RV.HYPER` (*a, b, c*)   [Function]
>   Hypergeometric distribution when *b* objects out of *a* are drawn and *c* of the available objects are distinctive. Constraints: integer $a > 0$, integer $b \leq a$, integer $c \leq a$, integer $x \geq 0$.

`PDF.LOG` (*x, p*)   [Function]
`RV.LOG` (*p*)   [Function]
>   Logarithmic distribution with probability parameter *p*. Constraints: $0 \leq p < 1$, $x \geq 1$.

`PDF.NEGBIN` (*x, n, p*)   [Function]
`CDF.NEGBIN` (*x, n, p*)   [Function]
`RV.NEGBIN` (*n, p*)   [Function]
>   Negative binomial distribution with number of successes parameter *n* and probability of success parameter *p*. Constraints: integer $n \geq 0$, $0 < p \leq 1$, integer $x \geq 1$.

`PDF.POISSON` (*x, mu*)   [Function]
`CDF.POISSON` (*x, mu*)   [Function]
`RV.POISSON` (*mu*)   [Function]
>   Poisson distribution with mean *mu*. Constraints: $mu > 0$, integer $x \geq 0$.

7.8 Operator Precedence

The following table describes operator precedence. Smaller-numbered levels in the table have higher precedence. Within a level, operations are always performed from left to right. The first occurrence of '-' represents unary negation, the second binary subtraction.

1. ( )
2. **
3. -
4. * /
5. + -
6. EQ GE GT LE LT NE
7. AND NOT OR

8 Data Input and Output

Data are the focus of the PSPP language. Each datum belongs to a *case* (also called an *observation*). Each case represents an individual or "experimental unit". For example, in the results of a survey, the names of the respondents, their sex, age, etc. and their responses are all data and the data pertaining to single respondent is a case. This chapter examines the PSPP commands for defining variables and reading and writing data. There are alternative commands to read data from predefined sources such as system files or databases (See Section 9.3 [GET], page 82.)

> **Note:** These commands tell PSPP how to read data, but the data will not actually be read until a procedure is executed.

8.1 BEGIN DATA

    BEGIN DATA.
    ...
    END DATA.

BEGIN DATA and END DATA can be used to embed raw ASCII data in a PSPP syntax file. DATA LIST or another input procedure must be used before BEGIN DATA (see Section 8.5 [DATA LIST], page 66). BEGIN DATA and END DATA must be used together. END DATA must appear by itself on a single line, with no leading white space and exactly one space between the words END and DATA, like this:

    END DATA.

8.2 CLOSE FILE HANDLE

    CLOSE FILE HANDLE *handle_name*.

CLOSE FILE HANDLE disassociates the name of a file handle with a given file. The only specification is the name of the handle to close. Afterward FILE HANDLE.

The file named INLINE, which represents data entered between BEGIN DATA and END DATA, cannot be closed. Attempts to close it with CLOSE FILE HANDLE have no effect.

CLOSE FILE HANDLE is a PSPP extension.

8.3 DATAFILE ATTRIBUTE

    DATAFILE ATTRIBUTE
        ATTRIBUTE=*name*('*value*') [*name*('*value*')]...
        ATTRIBUTE=*name*[*index*]('*value*') [*name*[*index*]('*value*')]...
        DELETE=*name* [*name*]...
        DELETE=*name*[*index*] [*name*[*index*]]...

DATAFILE ATTRIBUTE adds, modifies, or removes user-defined attributes associated with the active dataset. Custom data file attributes are not interpreted by PSPP, but they are saved as part of system files and may be used by other software that reads them.

Use the ATTRIBUTE subcommand to add or modify a custom data file attribute. Specify the name of the attribute as an identifier (see Section 6.1 [Tokens], page 28), followed by the desired value, in parentheses, as a quoted string. Attribute names that begin with $

are reserved for PSPP's internal use, and attribute names that begin with @ or $@ are not displayed by most PSPP commands that display other attributes. Other attribute names are not treated specially.

Attributes may also be organized into arrays. To assign to an array element, add an integer array index enclosed in square brackets ([and]) between the attribute name and value. Array indexes start at 1, not 0. An attribute array that has a single element (number 1) is not distinguished from a non-array attribute.

Use the DELETE subcommand to delete an attribute. Specify an attribute name by itself to delete an entire attribute, including all array elements for attribute arrays. Specify an attribute name followed by an array index in square brackets to delete a single element of an attribute array. In the latter case, all the array elements numbered higher than the deleted element are shifted down, filling the vacated position.

To associate custom attributes with particular variables, instead of with the entire active dataset, use VARIABLE ATTRIBUTE (see Section 11.14 [VARIABLE ATTRIBUTE], page 106) instead.

DATAFILE ATTRIBUTE takes effect immediately. It is not affected by conditional and looping structures such as DO IF or LOOP.

8.4 DATASET commands

```
DATASET NAME name [WINDOW={ASIS,FRONT}].
DATASET ACTIVATE name [WINDOW={ASIS,FRONT}].
DATASET COPY name [WINDOW={MINIMIZED,HIDDEN,FRONT}].
DATASET DECLARE name [WINDOW={MINIMIZED,HIDDEN,FRONT}].
DATASET CLOSE {name,*,ALL}.
DATASET DISPLAY.
```

The DATASET commands simplify use of multiple datasets within a PSPP session. They allow datasets to be created and destroyed. At any given time, most PSPP commands work with a single dataset, called the active dataset.

The DATASET NAME command gives the active dataset the specified name, or if it already had a name, it renames it. If another dataset already had the given name, that dataset is deleted.

The DATASET ACTIVATE command selects the named dataset, which must already exist, as the active dataset. Before switching the active dataset, any pending transformations are executed, as if EXECUTE had been specified. If the active dataset is unnamed before switching, then it is deleted and becomes unavailable after switching.

The DATASET COPY command creates a new dataset with the specified name, whose contents are a copy of the active dataset. Any pending transformations are executed, as if EXECUTE had been specified, before making the copy. If a dataset with the given name already exists, it is replaced. If the name is the name of the active dataset, then the active dataset becomes unnamed.

The DATASET DECLARE command creates a new dataset that is initially "empty," that is, it has no dictionary or data. If a dataset with the given name already exists, this has no effect. The new dataset can be used with commands that support output to a dataset, e.g. AGGREGATE (see Section 12.1 [AGGREGATE], page 110).

The DATASET CLOSE command deletes a dataset. If the active dataset is specified by name, or if '*' is specified, then the active dataset becomes unnamed. If a different dataset is specified by name, then it is deleted and becomes unavailable. Specifying ALL deletes all datasets except for the active dataset, which becomes unnamed.

The DATASET DISPLAY command lists all the currently defined datasets.

Many DATASET commands accept an optional WINDOW subcommand. In the PSPPIRE GUI, the value given for this subcommand influences how the dataset's window is displayed. Outside the GUI, the WINDOW subcommand has no effect. The valid values are:

ASIS   Do not change how the window is displayed. This is the default for DATASET NAME and DATASET ACTIVATE.

FRONT   Raise the dataset's window to the top. Make it the default dataset for running syntax.

MINIMIZED
            Display the window "minimized" to an icon. Prefer other datasets for running syntax. This is the default for DATASET COPY and DATASET DECLARE.

HIDDEN   Hide the dataset's window. Prefer other datasets for running syntax.

8.5 DATA LIST

Used to read text or binary data, DATA LIST is the most fundamental data-reading command. Even the more sophisticated input methods use DATA LIST commands as a building block. Understanding DATA LIST is important to understanding how to use PSPP to read your data files.

There are two major variants of DATA LIST, which are fixed format and free format. In addition, free format has a minor variant, list format, which is discussed in terms of its differences from vanilla free format.

Each form of DATA LIST is described in detail below.

See Section 9.4 [GET DATA], page 83, for a command that offers a few enhancements over DATA LIST and that may be substituted for DATA LIST in many situations.

8.5.1 DATA LIST FIXED

```
DATA LIST [FIXED]
    {TABLE,NOTABLE}
    [FILE='file_name' [ENCODING='encoding']]
    [RECORDS=record_count]
    [END=end_var]
    [SKIP=record_count]
    /[line_no] var_spec...
```

where each var_spec takes one of the forms

```
    var_list start-end [type_spec]
    var_list (fortran_spec)
```

DATA LIST FIXED is used to read data files that have values at fixed positions on each line of single-line or multiline records. The keyword FIXED is optional.

The `FILE` subcommand must be used if input is to be taken from an external file. It may be used to specify a file name as a string or a file handle (see Section 6.9 [File Handles], page 44). If the `FILE` subcommand is not used, then input is assumed to be specified within the command file using `BEGIN DATA...END DATA` (see Section 8.1 [BEGIN DATA], page 64). The `ENCODING` subcommand may only be used if the `FILE` subcommand is also used. It specifies the character encoding of the file. See Section 16.16 [INSERT], page 155, for information on supported encodings.

The optional `RECORDS` subcommand, which takes a single integer as an argument, is used to specify the number of lines per record. If `RECORDS` is not specified, then the number of lines per record is calculated from the list of variable specifications later in `DATA LIST`.

The `END` subcommand is only useful in conjunction with `INPUT PROGRAM`. See Section 8.9 [INPUT PROGRAM], page 73, for details.

The optional `SKIP` subcommand specifies a number of records to skip at the beginning of an input file. It can be used to skip over a row that contains variable names, for example.

`DATA LIST` can optionally output a table describing how the data file will be read. The `TABLE` subcommand enables this output, and `NOTABLE` disables it. The default is to output the table.

The list of variables to be read from the data list must come last. Each line in the data record is introduced by a slash ('/'). Optionally, a line number may follow the slash. Following, any number of variable specifications may be present.

Each variable specification consists of a list of variable names followed by a description of their location on the input line. Sets of variables may be specified using the `DATA LIST TO` convention (see Section 6.7.3 [Sets of Variables], page 34). There are two ways to specify the location of the variable on the line: columnar style and FORTRAN style.

In columnar style, the starting column and ending column for the field are specified after the variable name, separated by a dash ('-'). For instance, the third through fifth columns on a line would be specified '3-5'. By default, variables are considered to be in 'F' format (see Section 6.7.4 [Input and Output Formats], page 34). (This default can be changed; see Section 16.20 [SET], page 157 for more information.)

In columnar style, to use a variable format other than the default, specify the format type in parentheses after the column numbers. For instance, for alphanumeric 'A' format, use '(A)'.

In addition, implied decimal places can be specified in parentheses after the column numbers. As an example, suppose that a data file has a field in which the characters '1234' should be interpreted as having the value 12.34. Then this field has two implied decimal places, and the corresponding specification would be '(2)'. If a field that has implied decimal places contains a decimal point, then the implied decimal places are not applied.

Changing the variable format and adding implied decimal places can be done together; for instance, '(N,5)'.

When using columnar style, the input and output width of each variable is computed from the field width. The field width must be evenly divisible into the number of variables specified.

FORTRAN style is an altogether different approach to specifying field locations. With this approach, a list of variable input format specifications, separated by commas, are placed after the variable names inside parentheses. Each format specifier advances as many characters into the input line as it uses.

Implied decimal places also exist in FORTRAN style. A format specification with d decimal places also has d implied decimal places.

In addition to the standard format specifiers (see Section 6.7.4 [Input and Output Formats], page 34), FORTRAN style defines some extensions:

X   Advance the current column on this line by one character position.

Tx Set the current column on this line to column x, with column numbers considered to begin with 1 at the left margin.

NEWRECx Skip forward x lines in the current record, resetting the active column to the left margin.

Repeat count
           Any format specifier may be preceded by a number. This causes the action of that format specifier to be repeated the specified number of times.

($spec1$, ..., $specN$)
           Group the given specifiers together. This is most useful when preceded by a repeat count. Groups may be nested arbitrarily.

FORTRAN and columnar styles may be freely intermixed. Columnar style leaves the active column immediately after the ending column specified. Record motion using NEWREC in FORTRAN style also applies to later FORTRAN and columnar specifiers.

Examples

1.

```
DATA LIST TABLE /NAME 1-10 (A) INFO1 TO INFO3 12-17 (1).

BEGIN DATA.
John Smith 102311
Bob Arnold 122015
Bill Yates  918 6
END DATA.
```

Defines the following variables:

- NAME, a 10-character-wide string variable, in columns 1 through 10.
- INFO1, a numeric variable, in columns 12 through 13.
- INFO2, a numeric variable, in columns 14 through 15.
- INFO3, a numeric variable, in columns 16 through 17.

The BEGIN DATA/END DATA commands cause three cases to be defined:

Case	NAME	INFO1	INFO2	INFO3
1	John Smith	10	23	11
2	Bob Arnold	12	20	15
3	Bill Yates	9	18	6

The TABLE keyword causes PSPP to print out a table describing the four variables defined.

2.

```
DAT LIS FIL="survey.dat"
        /ID 1-5 NAME 7-36 (A) SURNAME 38-67 (A) MINITIAL 69 (A)
        /Q01 TO Q50 7-56
        /.
```

Defines the following variables:

- ID, a numeric variable, in columns 1-5 of the first record.

- NAME, a 30-character string variable, in columns 7-36 of the first record.

- SURNAME, a 30-character string variable, in columns 38-67 of the first record.

- MINITIAL, a 1-character string variable, in column 69 of the first record.

- Fifty variables Q01, Q02, Q03, ..., Q49, Q50, all numeric, Q01 in column 7, Q02 in column 8, ..., Q49 in column 55, Q50 in column 56, all in the second record.

Cases are separated by a blank record.

Data is read from file survey.dat in the current directory.

This example shows keywords abbreviated to their first 3 letters.

8.5.2 DATA LIST FREE

```
DATA LIST FREE
        [({TAB,'c'}, ...)]
        [{NOTABLE,TABLE}]
        [FILE='file_name' [ENCODING='encoding']]
        [SKIP=record_cnt]
        /var_spec...
```

where each var_spec takes one of the forms

var_list [(type_spec)]
var_list *

In free format, the input data is, by default, structured as a series of fields separated by spaces, tabs, commas, or line breaks. Each field's content may be unquoted, or it may be quoted with a pairs of apostrophes (' ') or double quotes (" "). Unquoted white space separates fields but is not part of any field. Any mix of spaces, tabs, and line breaks is equivalent to a single space for the purpose of separating fields, but consecutive commas will skip a field.

Alternatively, delimiters can be specified explicitly, as a parenthesized, comma-separated list of single-character strings immediately following FREE. The word TAB may also be used to specify a tab character as a delimiter. When delimiters are specified explicitly, only the given characters, plus line breaks, separate fields. Furthermore, leading spaces at the beginnings of fields are not trimmed, consecutive delimiters define empty fields, and no form of quoting is allowed.

The NOTABLE and TABLE subcommands are as in DATA LIST FIXED above. NOTABLE is the default.

The FILE, SKIP, and ENCODING subcommands are as in DATA LIST FIXED above.

The variables to be parsed are given as a single list of variable names. This list must be introduced by a single slash ('/'). The set of variable names may contain format specifications in parentheses (see Section 6.7.4 [Input and Output Formats], page 34). Format specifications apply to all variables back to the previous parenthesized format specification.

In addition, an asterisk may be used to indicate that all variables preceding it are to have input/output format 'F8.0'.

Specified field widths are ignored on input, although all normal limits on field width apply, but they are honored on output.

8.5.3 DATA LIST LIST

```
DATA LIST LIST
    [({TAB,'c'}, ...)]
    [{NOTABLE,TABLE}]
    [FILE='file_name' [ENCODING='encoding']]
    [SKIP=record_count]
    /var_spec...
```

where each *var_spec* takes one of the forms

```
    var_list [(type_spec)]
    var_list *
```

With one exception, **DATA LIST LIST** is syntactically and semantically equivalent to **DATA LIST FREE**. The exception is that each input line is expected to correspond to exactly one input record. If more or fewer fields are found on an input line than expected, an appropriate diagnostic is issued.

8.6 END CASE

```
END CASE.
```

END CASE is used only within **INPUT PROGRAM** to output the current case. See Section 8.9 [INPUT PROGRAM], page 73, for details.

8.7 END FILE

```
END FILE.
```

END FILE is used only within **INPUT PROGRAM** to terminate the current input program. See Section 8.9 [INPUT PROGRAM], page 73.

8.8 FILE HANDLE

For text files:

```
FILE HANDLE handle_name
        /NAME='file_name
        [/MODE=CHARACTER]
        [/ENDS={CR,CRLF}]
        /TABWIDTH=tab_width
        [ENCODING='encoding']
```

For binary files in native encoding with fixed-length records:
> FILE HANDLE *handle_name*
> 　　/NAME='*file_name*'
> 　　/MODE=IMAGE
> 　　[/LRECL=*rec_len*]
> 　　[ENCODING='*encoding*']

For binary files in native encoding with variable-length records:
> FILE HANDLE *handle_name*
> 　　/NAME='*file_name*'
> 　　/MODE=BINARY
> 　　[/LRECL=*rec_len*]
> 　　[ENCODING='*encoding*']

For binary files encoded in EBCDIC:
> FILE HANDLE *handle_name*
> 　　/NAME='*file_name*'
> 　　/MODE=360
> 　　/RECFORM={FIXED,VARIABLE,SPANNED}
> 　　[/LRECL=*rec_len*]
> 　　[ENCODING='*encoding*']

Use **FILE HANDLE** to associate a file handle name with a file and its attributes, so that later commands can refer to the file by its handle name. Names of text files can be specified directly on commands that access files, so that **FILE HANDLE** is only needed when a file is not an ordinary file containing lines of text. However, **FILE HANDLE** may be used even for text files, and it may be easier to specify a file's name once and later refer to it by an abstract handle.

Specify the file handle name as the identifier immediately following the **FILE HANDLE** command name. The identifier INLINE is reserved for representing data embedded in the syntax file (see Section 8.1 [BEGIN DATA], page 64) The file handle name must not already have been used in a previous invocation of **FILE HANDLE**, unless it has been closed by an intervening command (see Section 8.2 [CLOSE FILE HANDLE], page 64).

The effect and syntax of **FILE HANDLE** depends on the selected MODE:

- In CHARACTER mode, the default, the data file is read as a text file. Each text line is read as one record.

  In CHARACTER mode only, tabs are expanded to spaces by input programs, except by **DATA LIST FREE** with explicitly specified delimiters. Each tab is 4 characters wide by default, but TABWIDTH (a PSPP extension) may be used to specify an alternate width. Use a TABWIDTH of 0 to suppress tab expansion.

  A file written in CHARACTER mode by default uses the line ends of the system on which PSPP is running, that is, on Windows, the default is CR LF line ends, and on other systems the default is LF only. Specify ENDS as CR or CRLF to override the default. PSPP reads files using either convention on any kind of system, regardless of ENDS.

- In IMAGE mode, the data file is treated as a series of fixed-length binary records. LRECL should be used to specify the record length in bytes, with a default of 1024. On input, it is an error if an IMAGE file's length is not a integer multiple of the record length. On output, each record is padded with spaces or truncated, if necessary, to make it exactly the correct length.

- In BINARY mode, the data file is treated as a series of variable-length binary records. LRECL may be specified, but its value is ignored. The data for each record is both preceded and followed by a 32-bit signed integer in little-endian byte order that specifies the length of the record. (This redundancy permits records in these files to be efficiently read in reverse order, although PSPP always reads them in forward order.) The length does not include either integer.

- Mode 360 reads and writes files in formats first used for tapes in the 1960s on IBM mainframe operating systems and still supported today by the modern successors of those operating systems. For more information, see *OS/400 Tape and Diskette Device Programming*, available on IBM's website.

  Alphanumeric data in mode 360 files are encoded in EBCDIC. PSPP translates EBCDIC to or from the host's native format as necessary on input or output, using an ASCII/EBCDIC translation that is one-to-one, so that a "round trip" from ASCII to EBCDIC back to ASCII, or vice versa, always yields exactly the original data.

  The RECFORM subcommand is required in mode 360. The precise file format depends on its setting:

  F
  FIXED   This record format is equivalent to IMAGE mode, except for EBCDIC translation.

             IBM documentation calls this *F (fixed-length, deblocked) format.

  V
  VARIABLE
             The file comprises a sequence of zero or more variable-length blocks. Each block begins with a 4-byte *block descriptor word* (BDW). The first two bytes of the BDW are an unsigned integer in big-endian byte order that specifies the length of the block, including the BDW itself. The other two bytes of the BDW are ignored on input and written as zeros on output.

             Following the BDW, the remainder of each block is a sequence of one or more variable-length records, each of which in turn begins with a 4-byte *record descriptor word* (RDW) that has the same format as the BDW. Following the RDW, the remainder of each record is the record data.

             The maximum length of a record in VARIABLE mode is 65,527 bytes: 65,535 bytes (the maximum value of a 16-bit unsigned integer), minus 4 bytes for the BDW, minus 4 bytes for the RDW.

             In mode VARIABLE, LRECL specifies a maximum, not a fixed, record length, in bytes. The default is 8,192.

             IBM documentation calls this *VB (variable-length, blocked, unspanned) format.

VS
SPANNED

> The file format is like that of VARIABLE mode, except that logical records may be split among multiple physical records (called *segments*) or blocks. In SPANNED mode, the third byte of each RDW is called the segment control character (SCC). Odd SCC values cause the segment to be appended to a record buffer maintained in memory; even values also append the segment and then flush its contents to the input procedure. Canonically, SCC value 0 designates a record not spanned among multiple segments, and values 1 through 3 designate the first segment, the last segment, or an intermediate segment, respectively, within a multi-segment record. The record buffer is also flushed at end of file regardless of the final record's SCC.

> The maximum length of a logical record in VARIABLE mode is limited only by memory available to PSPP. Segments are limited to 65,527 bytes, as in VARIABLE mode.

> This format is similar to what IBM documentation call *VS (variable-length, deblocked, spanned) format.

In mode 360, fields of type A that extend beyond the end of a record read from disk are padded with spaces in the host's native character set, which are then translated from EBCDIC to the native character set. Thus, when the host's native character set is based on ASCII, these fields are effectively padded with character X'80'. This wart is implemented for compatibility.

The NAME subcommand specifies the name of the file associated with the handle. It is required in all modes but SCRATCH mode, in which its use is forbidden.

The ENCODING subcommand specifies the encoding of text in the file. For reading text files in CHARACTER mode, all of the forms described for ENCODING on the INSERT command are supported (see Section 16.16 [INSERT], page 155). For reading in other file-based modes, encoding autodetection is not supported; if the specified encoding requests autodetection then the default encoding will be used. This is also true when a file handle is used for writing a file in any mode.

8.9 INPUT PROGRAM

INPUT PROGRAM.
... input commands ...
END INPUT PROGRAM.

INPUT PROGRAM...END INPUT PROGRAM specifies a complex input program. By placing data input commands within INPUT PROGRAM, PSPP programs can take advantage of more complex file structures than available with only DATA LIST.

The first sort of extended input program is to simply put multiple DATA LIST commands within the INPUT PROGRAM. This will cause all of the data files to be read in parallel. Input will stop when end of file is reached on any of the data files.

Transformations, such as conditional and looping constructs, can also be included within INPUT PROGRAM. These can be used to combine input from several data files in more complex ways. However, input will still stop when end of file is reached on any of the data files.

To prevent `INPUT PROGRAM` from terminating at the first end of file, use the `END` subcommand on `DATA LIST`. This subcommand takes a variable name, which should be a numeric scratch variable (see Section 6.7.5 [Scratch Variables], page 43). (It need not be a scratch variable but otherwise the results can be surprising.) The value of this variable is set to 0 when reading the data file, or 1 when end of file is encountered.

Two additional commands are useful in conjunction with `INPUT PROGRAM`. `END CASE` is the first. Normally each loop through the `INPUT PROGRAM` structure produces one case. `END CASE` controls exactly when cases are output. When `END CASE` is used, looping from the end of `INPUT PROGRAM` to the beginning does not cause a case to be output.

`END FILE` is the second. When the `END` subcommand is used on `DATA LIST`, there is no way for the `INPUT PROGRAM` construct to stop looping, so an infinite loop results. `END FILE`, when executed, stops the flow of input data and passes out of the `INPUT PROGRAM` structure.

`INPUT PROGRAM` must contain at least one `DATA LIST` or `END FILE` command.

All this is very confusing. A few examples should help to clarify.

```
INPUT PROGRAM.
        DATA LIST NOTABLE FILE='a.data'/X 1-10.
        DATA LIST NOTABLE FILE='b.data'/Y 1-10.
END INPUT PROGRAM.
LIST.
```

The example above reads variable X from file `a.data` and variable Y from file `b.data`. If one file is shorter than the other then the extra data in the longer file is ignored.

```
INPUT PROGRAM.
        NUMERIC #A #B.

        DO IF NOT #A.
                DATA LIST NOTABLE END=#A FILE='a.data'/X 1-10.
        END IF.
        DO IF NOT #B.
                DATA LIST NOTABLE END=#B FILE='b.data'/Y 1-10.
        END IF.
        DO IF #A AND #B.
                END FILE.
        END IF.
        END CASE.
END INPUT PROGRAM.
LIST.
```

The above example reads variable X from `a.data` and variable Y from `b.data`. If one file is shorter than the other then the missing field is set to the system-missing value alongside the present value for the remaining length of the longer file.

```
INPUT PROGRAM.
        NUMERIC #A #B.

        DO IF #A.
                DATA LIST NOTABLE END=#B FILE='b.data'/X 1-10.
                DO IF #B.
```

```
                              END FILE.
                   ELSE.
                              END CASE.
                   END IF.
         ELSE.
                   DATA LIST NOTABLE END=#A FILE='a.data'/X 1-10.
                   DO IF NOT #A.
                              END CASE.
                   END IF.
         END IF.
   END INPUT PROGRAM.
   LIST.
```

The above example reads data from file a.data, then from b.data, and concatenates them into a single active dataset.

```
   INPUT PROGRAM.
         NUMERIC #EOF.

         LOOP IF NOT #EOF.
                   DATA LIST NOTABLE END=#EOF FILE='a.data'/X 1-10.
                   DO IF NOT #EOF.
                              END CASE.
                   END IF.
         END LOOP.

         COMPUTE #EOF = 0.
         LOOP IF NOT #EOF.
                   DATA LIST NOTABLE END=#EOF FILE='b.data'/X 1-10.
                   DO IF NOT #EOF.
                              END CASE.
                   END IF.
         END LOOP.

         END FILE.
   END INPUT PROGRAM.
   LIST.
```

The above example does the same thing as the previous example, in a different way.

```
   INPUT PROGRAM.
         LOOP #I=1 TO 50.
                   COMPUTE X=UNIFORM(10).
                   END CASE.
         END LOOP.
         END FILE.
   END INPUT PROGRAM.
   LIST/FORMAT=NUMBERED.
```

The above example causes an active dataset to be created consisting of 50 random variates between 0 and 10.

8.10 LIST

> LIST
>> /VARIABLES=*var_list*
>> /CASES=FROM *start_index* TO *end_index* BY *incr_index*
>> /FORMAT={UNNUMBERED,NUMBERED} {WRAP,SINGLE}

The LIST procedure prints the values of specified variables to the listing file.

The VARIABLES subcommand specifies the variables whose values are to be printed. Keyword VARIABLES is optional. If VARIABLES subcommand is not specified then all variables in the active dataset are printed.

The CASES subcommand can be used to specify a subset of cases to be printed. Specify FROM and the case number of the first case to print, TO and the case number of the last case to print, and BY and the number of cases to advance between printing cases, or any subset of those settings. If CASES is not specified then all cases are printed.

The FORMAT subcommand can be used to change the output format. NUMBERED will print case numbers along with each case; UNNUMBERED, the default, causes the case numbers to be omitted. The WRAP and SINGLE settings are currently not used.

Case numbers start from 1. They are counted after all transformations have been considered.

LIST attempts to fit all the values on a single line. If needed to make them fit, variable names are displayed vertically. If values cannot fit on a single line, then a multi-line format will be used.

LIST is a procedure. It causes the data to be read.

8.11 NEW FILE

> NEW FILE.

NEW FILE command clears the dictionary and data from the current active dataset.

8.12 PRINT

> PRINT
>> [OUTFILE='*file_name*']
>> [RECORDS=*n_lines*]
>> [{NOTABLE,TABLE}]
>> [ENCODING='*encoding*']
>> [/[*line_no*] *arg*...]

>> *arg* takes one of the following forms:
>>> '*string*' [*start*]
>>> *var_list* start-end [*type_spec*]
>>> *var_list* (*fortran_spec*)
>>> *var_list* *

The PRINT transformation writes variable data to the listing file or an output file. PRINT is executed when a procedure causes the data to be read. Follow PRINT by EXECUTE to print variable data without invoking a procedure (see Section 16.11 [EXECUTE], page 154).

All `PRINT` subcommands are optional. If no strings or variables are specified, `PRINT` outputs a single blank line.

The `OUTFILE` subcommand specifies the file to receive the output. The file may be a file name as a string or a file handle (see Section 6.9 [File Handles], page 44). If `OUTFILE` is not present then output will be sent to PSPP's output listing file. When `OUTFILE` is present, a space is inserted at beginning of each output line, even lines that otherwise would be blank.

The `ENCODING` subcommand may only be used if the `OUTFILE` subcommand is also used. It specifies the character encoding of the file. See Section 16.16 [INSERT], page 155, for information on supported encodings.

The `RECORDS` subcommand specifies the number of lines to be output. The number of lines may optionally be surrounded by parentheses.

`TABLE` will cause the `PRINT` command to output a table to the listing file that describes what it will print to the output file. `NOTABLE`, the default, suppresses this output table.

Introduce the strings and variables to be printed with a slash ('/'). Optionally, the slash may be followed by a number indicating which output line will be specified. In the absence of this line number, the next line number will be specified. Multiple lines may be specified using multiple slashes with the intended output for a line following its respective slash.

Literal strings may be printed. Specify the string itself. Optionally the string may be followed by a column number, specifying the column on the line where the string should start. Otherwise, the string will be printed at the current position on the line.

Variables to be printed can be specified in the same ways as available for `DATA LIST FIXED` (see Section 8.5.1 [DATA LIST FIXED], page 66). In addition, a variable list may be followed by an asterisk ('*'), which indicates that the variables should be printed in their dictionary print formats, separated by spaces. A variable list followed by a slash or the end of command will be interpreted the same way.

If a FORTRAN type specification is used to move backwards on the current line, then text is written at that point on the line, the line will be truncated to that length, although additional text being added will again extend the line to that length.

8.13 PRINT EJECT

```
PRINT EJECT
        OUTFILE='file_name'
        RECORDS=n_lines
        {NOTABLE,TABLE}
        /[line_no] arg...

arg takes one of the following forms:
        'string' [start-end]
        var_list start-end [type_spec]
        var_list (fortran_spec)
        var_list *
```

`PRINT EJECT` advances to the beginning of a new output page in the listing file or output file. It can also output data in the same way as `PRINT`.

All `PRINT EJECT` subcommands are optional.

Without OUTFILE, PRINT EJECT ejects the current page in the listing file, then it produces other output, if any is specified.

With OUTFILE, PRINT EJECT writes its output to the specified file. The first line of output is written with '1' inserted in the first column. Commonly, this is the only line of output. If additional lines of output are specified, these additional lines are written with a space inserted in the first column, as with PRINT.

See Section 8.12 [PRINT], page 76, for more information on syntax and usage.

8.14 PRINT SPACE

> PRINT SPACE [OUTFILE='file_name'] [ENCODING='encoding'] [n_lines].

PRINT SPACE prints one or more blank lines to an output file.

The OUTFILE subcommand is optional. It may be used to direct output to a file specified by file name as a string or file handle (see Section 6.9 [File Handles], page 44). If OUTFILE is not specified then output will be directed to the listing file.

The ENCODING subcommand may only be used if OUTFILE is also used. It specifies the character encoding of the file. See Section 16.16 [INSERT], page 155, for information on supported encodings.

n_lines is also optional. If present, it is an expression (see Chapter 7 [Expressions], page 46) specifying the number of blank lines to be printed. The expression must evaluate to a nonnegative value.

8.15 REREAD

> REREAD [FILE=handle] [COLUMN=column] [ENCODING='encoding'].

The REREAD transformation allows the previous input line in a data file already processed by DATA LIST or another input command to be re-read for further processing.

The FILE subcommand, which is optional, is used to specify the file to have its line re-read. The file must be specified as the name of a file handle (see Section 6.9 [File Handles], page 44). If FILE is not specified then the last file specified on DATA LIST will be assumed (last file specified lexically, not in terms of flow-of-control).

By default, the line re-read is re-read in its entirety. With the COLUMN subcommand, a prefix of the line can be exempted from re-reading. Specify an expression (see Chapter 7 [Expressions], page 46) evaluating to the first column that should be included in the re-read line. Columns are numbered from 1 at the left margin.

The ENCODING subcommand may only be used if the FILE subcommand is also used. It specifies the character encoding of the file. See Section 16.16 [INSERT], page 155, for information on supported encodings.

Issuing REREAD multiple times will not back up in the data file. Instead, it will re-read the same line multiple times.

8.16 REPEATING DATA

> REPEATING DATA
>   /STARTS=start-end
>   /OCCURS=n_occurs

```
/FILE='file_name'
/LENGTH=length
/CONTINUED[=cont_start-cont_end]
/ID=id_start-id_end=id_var
/{TABLE,NOTABLE}
/DATA=var_spec...
```

where each *var_spec* takes one of the forms
var_list start-end [*type_spec*]
var_list (*fortran_spec*)

REPEATING DATA parses groups of data repeating in a uniform format, possibly with several groups on a single line. Each group of data corresponds with one case. **REPEATING DATA** may only be used within an **INPUT PROGRAM** structure (see Section 8.9 [INPUT PROGRAM], page 73). When used with **DATA LIST**, it can be used to parse groups of cases that share a subset of variables but differ in their other data.

The **STARTS** subcommand is required. Specify a range of columns, using literal numbers or numeric variable names. This range specifies the columns on the first line that are used to contain groups of data. The ending column is optional. If it is not specified, then the record width of the input file is used. For the inline file (see Section 8.1 [BEGIN DATA], page 64) this is 80 columns; for a file with fixed record widths it is the record width; for other files it is 1024 characters by default.

The **OCCURS** subcommand is required. It must be a number or the name of a numeric variable. Its value is the number of groups present in the current record.

The **DATA** subcommand is required. It must be the last subcommand specified. It is used to specify the data present within each repeating group. Column numbers are specified relative to the beginning of a group at column 1. Data is specified in the same way as with **DATA LIST FIXED** (see Section 8.5.1 [DATA LIST FIXED], page 66).

All other subcommands are optional.

FILE specifies the file to read, either a file name as a string or a file handle (see Section 6.9 [File Handles], page 44). If FILE is not present then the default is the last file handle used on **DATA LIST** (lexically, not in terms of flow of control).

By default **REPEATING DATA** will output a table describing how it will parse the input data. Specifying **NOTABLE** will disable this behavior; specifying TABLE will explicitly enable it.

The **LENGTH** subcommand specifies the length in characters of each group. If it is not present then length is inferred from the **DATA** subcommand. LENGTH can be a number or a variable name.

Normally all the data groups are expected to be present on a single line. Use the **CONTINUED** command to indicate that data can be continued onto additional lines. If data on continuation lines starts at the left margin and continues through the entire field width, no column specifications are necessary on **CONTINUED**. Otherwise, specify the possible range of columns in the same way as on STARTS.

When data groups are continued from line to line, it is easy for cases to get out of sync through careless hand editing. The **ID** subcommand allows a case identifier to be present on each line of repeating data groups. **REPEATING DATA** will check for the same identifier on each line and report mismatches. Specify the range of columns that the identifier will occupy, followed by an equals sign ('=') and the identifier variable name. The variable must already have been declared with NUMERIC or another command.

REPEATING DATA should be the last command given within an INPUT PROGRAM. It should not be enclosed within a LOOP structure (see Section 14.4 [LOOP], page 125). Use DATA LIST before, not after, REPEATING DATA.

8.17 WRITE

WRITE
        OUTFILE='file_name'
        RECORDS=n_lines
        {NOTABLE,TABLE}
        /[line_no] arg...

arg takes one of the following forms:
        'string' [start-end]
        var_list start-end [type_spec]
        var_list (fortran_spec)
        var_list *

WRITE writes text or binary data to an output file.

See Section 8.12 [PRINT], page 76, for more information on syntax and usage. PRINT and WRITE differ in only a few ways:

- WRITE uses write formats by default, whereas PRINT uses print formats.

- PRINT inserts a space between variables unless a format is explicitly specified, but WRITE never inserts space between variables in output.

- PRINT inserts a space at the beginning of each line that it writes to an output file (and PRINT EJECT inserts '1' at the beginning of each line that should begin a new page), but WRITE does not.

- PRINT outputs the system-missing value according to its specified output format, whereas WRITE outputs the system-missing value as a field filled with spaces. Binary formats are an exception.

9 System and Portable File I/O

The commands in this chapter read, write, and examine system files and portable files.

9.1 APPLY DICTIONARY

APPLY DICTIONARY FROM={'*file_name*',*file_handle*}.

`APPLY DICTIONARY` applies the variable labels, value labels, and missing values taken from a file to corresponding variables in the active dataset. In some cases it also updates the weighting variable.

Specify a system file or portable file's name, a data set name (see Section 6.7 [Datasets], page 32), or a file handle name (see Section 6.9 [File Handles], page 44). The dictionary in the file will be read, but it will not replace the active dataset's dictionary. The file's data will not be read.

Only variables with names that exist in both the active dataset and the system file are considered. Variables with the same name but different types (numeric, string) will cause an error message. Otherwise, the system file variables' attributes will replace those in their matching active dataset variables:

- If a system file variable has a variable label, then it will replace the variable label of the active dataset variable. If the system file variable does not have a variable label, then the active dataset variable's variable label, if any, will be retained.

- If the system file variable has custom attributes (see Section 11.14 [VARIABLE AT-TRIBUTE], page 106), then those attributes replace the active dataset variable's custom attributes. If the system file variable does not have custom attributes, then the active dataset variable's custom attributes, if any, will be retained.

- If the active dataset variable is numeric or short string, then value labels and missing values, if any, will be copied to the active dataset variable. If the system file variable does not have value labels or missing values, then those in the active dataset variable, if any, will not be disturbed.

In addition to properties of variables, some properties of the active file dictionary as a whole are updated:

- If the system file has custom attributes (see Section 8.3 [DATAFILE ATTRIBUTE], page 64), then those attributes replace the active dataset variable's custom attributes.

- If the active dataset has a weighting variable (see Section 13.7 [WEIGHT], page 123), and the system file does not, or if the weighting variable in the system file does not exist in the active dataset, then the active dataset weighting variable, if any, is retained. Otherwise, the weighting variable in the system file becomes the active dataset weighting variable.

`APPLY DICTIONARY` takes effect immediately. It does not read the active dataset. The system file is not modified.

9.2 EXPORT

EXPORT
        /OUTFILE='file_name'
        /UNSELECTED={RETAIN,DELETE}
        /DIGITS=n
        /DROP=var_list
        /KEEP=var_list
        /RENAME=(src_names=target_names)...
        /TYPE={COMM,TAPE}
        /MAP

The EXPORT procedure writes the active dataset's dictionary and data to a specified portable file.

By default, cases excluded with FILTER are written to the file. These can be excluded by specifying DELETE on the UNSELECTED subcommand. Specifying RETAIN makes the default explicit.

Portable files express real numbers in base 30. Integers are always expressed to the maximum precision needed to make them exact. Non-integers are, by default, expressed to the machine's maximum natural precision (approximately 15 decimal digits on many machines). If many numbers require this many digits, the portable file may significantly increase in size. As an alternative, the DIGITS subcommand may be used to specify the number of decimal digits of precision to write. DIGITS applies only to non-integers.

The OUTFILE subcommand, which is the only required subcommand, specifies the portable file to be written as a file name string or a file handle (see Section 6.9 [File Handles], page 44).

DROP, KEEP, and RENAME follow the same format as the SAVE procedure (see Section 9.6 [SAVE], page 89).

The TYPE subcommand specifies the character set for use in the portable file. Its value is currently not used.

The MAP subcommand is currently ignored.

EXPORT is a procedure. It causes the active dataset to be read.

9.3 GET

GET
        /FILE={'file_name',file_handle}
        /DROP=var_list
        /KEEP=var_list
        /RENAME=(src_names=target_names)...
        /ENCODING='encoding'

GET clears the current dictionary and active dataset and replaces them with the dictionary and data from a specified file.

The FILE subcommand is the only required subcommand. Specify the SPSS system file, SPSS/PC+ system file, or SPSS portable file to be read as a string file name or a file handle (see Section 6.9 [File Handles], page 44).

By default, all the variables in a file are read. The DROP subcommand can be used to specify a list of variables that are not to be read. By contrast, the KEEP subcommand can be used to specify variable that are to be read, with all other variables not read.

Normally variables in a file retain the names that they were saved under. Use the RENAME subcommand to change these names. Specify, within parentheses, a list of variable names followed by an equals sign ('=') and the names that they should be renamed to. Multiple parenthesized groups of variable names can be included on a single RENAME subcommand. Variables' names may be swapped using a RENAME subcommand of the form /RENAME=(A B=B A).

Alternate syntax for the RENAME subcommand allows the parentheses to be eliminated. When this is done, only a single variable may be renamed at once. For instance, /RENAME=A=B. This alternate syntax is deprecated.

DROP, KEEP, and RENAME are executed in left-to-right order. Each may be present any number of times. GET never modifies a file on disk. Only the active dataset read from the file is affected by these subcommands.

PSPP automatically detects the encoding of string data in the file, when possible. The character encoding of old SPSS system files cannot always be guessed correctly, and SPSS/PC+ system files do not include any indication of their encoding. Specify the ENCODING subcommand with an IANA character set name as its string argument to override the default. Use SYSFILE INFO to analyze the encodings that might be valid for a system file. The ENCODING subcommand is a PSPP extension.

GET does not cause the data to be read, only the dictionary. The data is read later, when a procedure is executed.

Use of GET to read a portable file is a PSPP extension.

9.4 GET DATA

    GET DATA
        /TYPE={GNM,ODS,PSQL,TXT}
        ...additional subcommands depending on TYPE...

The GET DATA command is used to read files and other data sources created by other applications. When this command is executed, the current dictionary and active dataset are replaced with variables and data read from the specified source.

The TYPE subcommand is mandatory and must be the first subcommand specified. It determines the type of the file or source to read. PSPP currently supports the following file types:

GNM   Spreadsheet files created by Gnumeric (http://gnumeric.org).

ODS   Spreadsheet files in OpenDocument format (http://opendocumentformat.org).

PSQL   Relations from PostgreSQL databases (http://postgresql.org).

TXT   Textual data files in columnar and delimited formats.

Each supported file type has additional subcommands, explained in separate sections below.

9.4.1 Spreadsheet Files

GET DATA /TYPE={GNM, ODS}
        /FILE={'file_name'}
        /SHEET={NAME 'sheet_name', INDEX n}
        /CELLRANGE={RANGE 'range', FULL}
        /READNAMES={ON, OFF}
        /ASSUMEDSTRWIDTH=n.

Gnumeric spreadsheets (http://gnumeric.org), and spreadsheets in OpenDocument format (http://libreplanet.org/wiki/Group:OpenDocument/Software) can be read using the GET DATA command. Use the TYPE subcommand to indicate the file's format. /TYPE=GNM indicates Gnumeric files, /TYPE=ODS indicates OpenDocument. The FILE subcommand is mandatory. Use it to specify the name file to be read. All other subcommands are optional.

The format of each variable is determined by the format of the spreadsheet cell containing the first datum for the variable. If this cell is of string (text) format, then the width of the variable is determined from the length of the string it contains, unless the ASSUMEDSTRWIDTH subcommand is given.

The SHEET subcommand specifies the sheet within the spreadsheet file to read. There are two forms of the SHEET subcommand. In the first form, /SHEET=name sheet_name, the string sheet_name is the name of the sheet to read. In the second form, /SHEET=index idx, idx is a integer which is the index of the sheet to read. The first sheet has the index 1. If the SHEET subcommand is omitted, then the command will read the first sheet in the file.

The CELLRANGE subcommand specifies the range of cells within the sheet to read. If the subcommand is given as /CELLRANGE=FULL, then the entire sheet is read. To read only part of a sheet, use the form /CELLRANGE=range 'top_left_cell:bottom_right_cell'. For example, the subcommand /CELLRANGE=range 'C3:P19' reads columns C–P, and rows 3–19 inclusive. If no CELLRANGE subcommand is given, then the entire sheet is read.

If /READNAMES=ON is specified, then the contents of cells of the first row are used as the names of the variables in which to store the data from subsequent rows. This is the default. If /READNAMES=OFF is used, then the variables receive automatically assigned names.

The ASSUMEDSTRWIDTH subcommand specifies the maximum width of string variables read from the file. If omitted, the default value is determined from the length of the string in the first spreadsheet cell for each variable.

9.4.2 Postgres Database Queries

GET DATA /TYPE=PSQL
        /CONNECT={connection info}
        /SQL={query}
        [/ASSUMEDSTRWIDTH=w]
        [/UNENCRYPTED]
        [/BSIZE=n].

The PSQL type is used to import data from a postgres database server. The server may be located locally or remotely. Variables are automatically created based on the table column names or the names specified in the SQL query. Postgres data types of high precision, will loose precision when imported into PSPP. Not all the postgres data types are able to be represented in PSPP. If a datum cannot be represented a warning will be issued and that datum will be set to SYSMIS.

The CONNECT subcommand is mandatory. It is a string specifying the parameters of the database server from which the data should be fetched. The format of the string is given in the postgres manual http://www.postgresql.org/docs/8.0/static/libpq.html#LIBPQ-CONNECT.

The SQL subcommand is mandatory. It must be a valid SQL string to retrieve data from the database.

The ASSUMEDSTRWIDTH subcommand specifies the maximum width of string variables read from the database. If omitted, the default value is determined from the length of the string in the first value read for each variable.

The UNENCRYPTED subcommand allows data to be retrieved over an insecure connection. If the connection is not encrypted, and the UNENCRYPTED subcommand is not given, then an error will occur. Whether or not the connection is encrypted depends upon the underlying psql library and the capabilities of the database server.

The BSIZE subcommand serves only to optimise the speed of data transfer. It specifies an upper limit on number of cases to fetch from the database at once. The default value is 4096. If your SQL statement fetches a large number of cases but only a small number of variables, then the data transfer may be faster if you increase this value. Conversely, if the number of variables is large, or if the machine on which PSPP is running has only a small amount of memory, then a smaller value will be better.

The following syntax is an example:

```
GET DATA /TYPE=PSQL
    /CONNECT='host=example.com port=5432 dbname=product user=fred passwd=xxxx'
    /SQL='select * from manufacturer'.
```

9.4.3 Textual Data Files

```
GET DATA /TYPE=TXT
    /FILE={'file_name',file_handle}
    [ENCODING='encoding']
    [/ARRANGEMENT={DELIMITED,FIXED}]
    [/FIRSTCASE={first_case}]
    [/IMPORTCASE={ALL,FIRST max_cases,PERCENT percent}]
    ...additional subcommands depending on ARRANGEMENT...
```

When TYPE=TXT is specified, GET DATA reads data in a delimited or fixed columnar format, much like DATA LIST (see Section 8.5 [DATA LIST], page 66).

The FILE subcommand is mandatory. Specify the file to be read as a string file name or (for textual data only) a file handle (see Section 6.9 [File Handles], page 44).

The ENCODING subcommand specifies the character encoding of the file to be read. See Section 16.16 [INSERT], page 155, for information on supported encodings.

The ARRANGEMENT subcommand determines the file's basic format. DELIMITED, the default setting, specifies that fields in the input data are separated by spaces, tabs, or other user-specified delimiters. FIXED specifies that fields in the input data appear at particular fixed column positions within records of a case.

By default, cases are read from the input file starting from the first line. To skip lines at the beginning of an input file, set `FIRSTCASE` to the number of the first line to read: 2 to skip the first line, 3 to skip the first two lines, and so on.

`IMPORTCASE` can be used to limit the number of cases read from the input file. With the default setting, ALL, all cases in the file are read. Specify FIRST max_cases to read at most max_cases cases from the file. Use `PERCENT percent` to read only percent percent, approximately, of the cases contained in the file. (The percentage is approximate, because there is no way to accurately count the number of cases in the file without reading the entire file. The number of cases in some kinds of unusual files cannot be estimated; PSPP will read all cases in such files.)

`FIRSTCASE` and `IMPORTCASE` may be used with delimited and fixed-format data. The remaining subcommands, which apply only to one of the two file arrangements, are described below.

9.4.3.1 Reading Delimited Data

```
GET DATA /TYPE=TXT
        /FILE={'file_name',file_handle}
        [/ARRANGEMENT={DELIMITED,FIXED}]
        [/FIRSTCASE={first_case}]
        [/IMPORTCASE={ALL,FIRST max_cases,PERCENT percent}]


        /DELIMITERS="delimiters"
        [/QUALIFIER="quotes" [/ESCAPE]]
        [/DELCASE={LINE,VARIABLES n_variables}]
        /VARIABLES=del_var1 [del_var2]...
    where each del_var takes the form:
        variable format
```

The GET DATA command with TYPE=TXT and ARRANGEMENT=DELIMITED reads input data from text files in delimited format, where fields are separated by a set of user-specified delimiters. Its capabilities are similar to those of DATA LIST FREE (see Section 8.5.2 [DATA LIST FREE], page 69), with a few enhancements.

The required `FILE` subcommand and optional `FIRSTCASE` and `IMPORTCASE` subcommands are described above (see Section 9.4.3 [GET DATA /TYPE=TXT], page 85).

`DELIMITERS`, which is required, specifies the set of characters that may separate fields. Each character in the string specified on `DELIMITERS` separates one field from the next. The end of a line also separates fields, regardless of `DELIMITERS`. Two consecutive delimiters in the input yield an empty field, as does a delimiter at the end of a line. A space character as a delimiter is an exception: consecutive spaces do not yield an empty field and neither does any number of spaces at the end of a line.

To use a tab as a delimiter, specify '\t' at the beginning of the `DELIMITERS` string. To use a backslash as a delimiter, specify '\\' as the first delimiter or, if a tab should also be a delimiter, immediately following '\t'. To read a data file in which each field appears on a separate line, specify the empty string for `DELIMITERS`.

The optional `QUALIFIER` subcommand names one or more characters that can be used to quote values within fields in the input. A field that begins with one of the specified quote characters ends at the next matching quote. Intervening delimiters become part of the field, instead of terminating it. The ability to specify more than one quote character is a PSPP extension.

By default, a character specified on `QUALIFIER` cannot itself be embedded within a field that it quotes, because the quote character always terminates the quoted field. With ESCAPE, however, a doubled quote character within a quoted field inserts a single instance of the quote into the field. For example, if ''' is specified on `QUALIFIER`, then without ESCAPE 'a''b' specifies a pair of fields that contain 'a' and 'b', but with ESCAPE it specifies a single field that contains 'a'b'. ESCAPE is a PSPP extension.

The `DELCASE` subcommand controls how data may be broken across lines in the data file. With LINE, the default setting, each line must contain all the data for exactly one case. For additional flexibility, to allow a single case to be split among lines or multiple cases to be contained on a single line, specify VARIABLES *n_variables*, where *n_variables* is the number of variables per case.

The `VARIABLES` subcommand is required and must be the last subcommand. Specify the name of each variable and its input format (see Section 6.7.4 [Input and Output Formats], page 34) in the order they should be read from the input file.

Examples

On a Unix-like system, the '/etc/passwd' file has a format similar to this:

```
root:$1$nyeSP5gD$pDq/:0:0:,,,:/root:/bin/bash
blp:$1$BrP/pFg4$g7OG:1000:1000:Ben Pfaff,,,:/home/blp:/bin/bash
john:$1$JBuq/Fioq$g4A:1001:1001:John Darrington,,,:/home/john:/bin/bash
jhs:$1$D3li4hPL$88X1:1002:1002:Jason Stover,,,:/home/jhs:/bin/csh
```

The following syntax reads a file in the format used by '/etc/passwd':

```
GET DATA /TYPE=TXT /FILE='/etc/passwd' /DELIMITERS=':'
        /VARIABLES=username A20
                password A40
                uid F10
                gid F10
                gecos A40
                home A40
                shell A40.
```

Consider the following data on used cars:

```
model   year    mileage price   type    age
Civic   2002    29883   15900   Si      2
Civic   2003    13415   15900   EX      1
Civic   1992    107000  3800    n/a     12
Accord  2002    26613   17900   EX      1
```

The following syntax can be used to read the used car data:

```
GET DATA /TYPE=TXT /FILE='cars.data' /DELIMITERS=' ' /FIRSTCASE=2
        /VARIABLES=model A8
                year F4
                mileage F6
```

```
price F5
type A4
age F2.
```

Consider the following information on animals in a pet store:

```
'Pet''s Name', "Age", "Color", "Date Received", "Price", "Height", "Type"
, (Years), , , (Dollars), ,
"Rover", 4.5, Brown, "12 Feb 2004", 80, '1''4"', "Dog"
"Charlie", , Gold, "5 Apr 2007", 12.3, "3"""", "Fish"
"Molly", 2, Black, "12 Dec 2006", 25, '5"', "Cat"
"Gilly", , White, "10 Apr 2007", 10, "3"""", "Guinea Pig"
```

The following syntax can be used to read the pet store data:

```
GET DATA /TYPE=TXT /FILE='pets.data' /DELIMITERS=', ' /QUALIFIER='''"' /ESCAPE
        /FIRSTCASE=3
        /VARIABLES=name A10
                   age F3.1
                   color A5
                   received EDATE10
                   price F5.2
                   height a5
                   type a10.
```

9.4.3.2 Reading Fixed Columnar Data

```
GET DATA /TYPE=TXT
        /FILE={'file_name',file_handle}
        [/ARRANGEMENT={DELIMITED,FIXED}]
        [/FIRSTCASE={first_case}]
        [/IMPORTCASE={ALL,FIRST max_cases,PERCENT percent}]

        [/FIXCASE=n]
        /VARIABLES fixed_var [fixed_var]...
            [/rec# fixed_var [fixed_var]...]...
    where each fixed_var takes the form:
        variable start-end format
```

The GET DATA command with TYPE=TXT and ARRANGEMENT=FIXED reads input data from text files in fixed format, where each field is located in particular fixed column positions within records of a case. Its capabilities are similar to those of DATA LIST FIXED (see Section 8.5.1 [DATA LIST FIXED], page 66), with a few enhancements.

The required FILE subcommand and optional FIRSTCASE and IMPORTCASE subcommands are described above (see Section 9.4.3 [GET DATA /TYPE=TXT], page 85).

The optional FIXCASE subcommand may be used to specify the positive integer number of input lines that make up each case. The default value is 1.

The VARIABLES subcommand, which is required, specifies the positions at which each variable can be found. For each variable, specify its name, followed by its start and end column separated by '-' (e.g. '0-9'), followed by an input format type (e.g. 'F') or a full format specification (e.g. 'DOLLAR12.2'). For this command, columns are numbered starting

from 0 at the left column. Introduce the variables in the second and later lines of a case by
a slash followed by the number of the line within the case, e.g. '/2' for the second line.

Examples

Consider the following data on used cars:

```
model   year   mileage price   type   age
Civic   2002   29883   15900   Si     2
Civic   2003   13415   15900   EX     1
Civic   1992   107000  3800    n/a    12
Accord  2002   26613   17900   EX     1
```

The following syntax can be used to read the used car data:

```
GET DATA /TYPE=TXT /FILE='cars.data' /ARRANGEMENT=FIXED /FIRSTCASE=2
         /VARIABLES=model 0-7 A
                    year 8-15 F
                    mileage 16-23 F
                    price 24-31 F
                    type 32-40 A
                    age 40-47 F.
```

9.5 IMPORT

```
IMPORT
    /FILE='file_name'
    /TYPE={COMM,TAPE}
    /DROP=var_list
    /KEEP=var_list
    /RENAME=(src_names=target_names)...
```

The IMPORT transformation clears the active dataset dictionary and data and replaces
them with a dictionary and data from a system file or portable file.

The FILE subcommand, which is the only required subcommand, specifies the portable
file to be read as a file name string or a file handle (see Section 6.9 [File Handles], page 44).

The TYPE subcommand is currently not used.

DROP, KEEP, and RENAME follow the syntax used by GET (see Section 9.3 [GET], page 82).

IMPORT does not cause the data to be read; only the dictionary. The data is read later,
when a procedure is executed.

Use of IMPORT to read a system file is a PSPP extension.

9.6 SAVE

```
SAVE
    /OUTFILE={'file_name',file_handle}
    /UNSELECTED={RETAIN,DELETE}
    /{UNCOMPRESSED,COMPRESSED,ZCOMPRESSED}
    /PERMISSIONS={WRITEABLE,READONLY}
    /DROP=var_list
    /KEEP=var_list
```

```
/VERSION=version
/RENAME=(src_names=target_names)...
/NAMES
/MAP
```

The **SAVE** procedure causes the dictionary and data in the active dataset to be written to a system file.

OUTFILE is the only required subcommand. Specify the system file to be written as a string file name or a file handle (see Section 6.9 [File Handles], page 44).

By default, cases excluded with FILTER are written to the system file. These can be excluded by specifying **DELETE** on the **UNSELECTED** subcommand. Specifying **RETAIN** makes the default explicit.

The **UNCOMPRESSED**, **COMPRESSED**, and **ZCOMPRESSED** subcommand determine the system file's compression level:

UNCOMPRESSED

> Data is not compressed. Each numeric value uses 8 bytes of disk space. Each string value uses one byte per column width, rounded up to a multiple of 8 bytes.

COMPRESSED

> Data is compressed with a simple algorithm. Each integer numeric value between −99 and 151, inclusive, or system missing value uses one byte of disk space. Each 8-byte segment of a string that consists only of spaces uses 1 byte. Any other numeric value or 8-byte string segment uses 9 bytes of disk space.

ZCOMPRESSED

> Data is compressed with the "deflate" compression algorithm specified in RFC 1951 (the same algorithm used by **gzip**). Files written with this compression level cannot be read by PSPP 0.8.1 or earlier or by SPSS 20 or earlier.

COMPRESSED is the default compression level. The SET command (see Section 16.20 [SET], page 157) can change this default.

The **PERMISSIONS** subcommand specifies permissions for the new system file. WRITE-ABLE, the default, creates the file with read and write permission. READONLY creates the file for read-only access.

By default, all the variables in the active dataset dictionary are written to the system file. The **DROP** subcommand can be used to specify a list of variables not to be written. In contrast, KEEP specifies variables to be written, with all variables not specified not written.

Normally variables are saved to a system file under the same names they have in the active dataset. Use the **RENAME** subcommand to change these names. Specify, within parentheses, a list of variable names followed by an equals sign ('=') and the names that they should be renamed to. Multiple parenthesized groups of variable names can be included on a single **RENAME** subcommand. Variables' names may be swapped using a **RENAME** subcommand of the form /RENAME=(A B=B A).

Alternate syntax for the **RENAME** subcommand allows the parentheses to be eliminated. When this is done, only a single variable may be renamed at once. For instance, /RENAME=A=B. This alternate syntax is deprecated.

DROP, KEEP, and RENAME are performed in left-to-right order. They each may be present any number of times. SAVE never modifies the active dataset. DROP, KEEP, and RENAME only affect the system file written to disk.

The VERSION subcommand specifies the version of the file format. Valid versions are 2 and 3. The default version is 3. In version 2 system files, variable names longer than 8 bytes will be truncated. The two versions are otherwise identical.

The NAMES and MAP subcommands are currently ignored.

SAVE causes the data to be read. It is a procedure.

9.7 SAVE TRANSLATE

```
SAVE TRANSLATE
      /OUTFILE={'file_name',file_handle}
      /TYPE={CSV,TAB}
      [/REPLACE]
      [/MISSING={IGNORE,RECODE}]

      [/DROP=var_list]
      [/KEEP=var_list]
      [/RENAME=(src_names=target_names)...]
      [/UNSELECTED={RETAIN,DELETE}]
      [/MAP]
```

...additional subcommands depending on TYPE...

The SAVE TRANSLATE command is used to save data into various formats understood by other applications.

The OUTFILE and TYPE subcommands are mandatory. OUTFILE specifies the file to be written, as a string file name or a file handle (see Section 6.9 [File Handles], page 44). TYPE determines the type of the file or source to read. It must be one of the following:

CSV   Comma-separated value format,

TAB   Tab-delimited format.

By default, SAVE TRANSLATE will not overwrite an existing file. Use REPLACE to force an existing file to be overwritten.

With MISSING=IGNORE, the default, SAVE TRANSLATE treats user-missing values as if they were not missing. Specify MISSING=RECODE to output numeric user-missing values like system-missing values and string user-missing values as all spaces.

By default, all the variables in the active dataset dictionary are saved to the system file, but DROP or KEEP can select a subset of variable to save. The RENAME subcommand can also be used to change the names under which variables are saved. UNSELECTED determines whether cases filtered out by the FILTER command are written to the output file. These subcommands have the same syntax and meaning as on the SAVE command (see Section 9.6 [SAVE], page 89).

Each supported file type has additional subcommands, explained in separate sections below.

SAVE TRANSLATE causes the data to be read. It is a procedure.

9.7.1 Writing Comma- and Tab-Separated Data Files

```
SAVE TRANSLATE
    /OUTFILE={'file_name',file_handle}
    /TYPE=CSV
    [/REPLACE]
    [/MISSING={IGNORE,RECODE}]

    [/DROP=var_list]
    [/KEEP=var_list]
    [/RENAME=(src_names=target_names)...]
    [/UNSELECTED={RETAIN,DELETE}]

    [/FIELDNAMES]
    [/CELLS={VALUES,LABELS}]
    [/TEXTOPTIONS DELIMITER='delimiter']
    [/TEXTOPTIONS QUALIFIER='qualifier']
    [/TEXTOPTIONS DECIMAL={DOT,COMMA}]
    [/TEXTOPTIONS FORMAT={PLAIN,VARIABLE}]
```

The SAVE TRANSLATE command with TYPE=CSV or TYPE=TAB writes data in a comma- or tab-separated value format similar to that described by RFC 4180. Each variable becomes one output column, and each case becomes one line of output. If FIELDNAMES is specified, an additional line at the top of the output file lists variable names.

The CELLS and TEXTOPTIONS FORMAT settings determine how values are written to the output file:

CELLS=VALUES FORMAT=PLAIN (the default settings)

Writes variables to the output in "plain" formats that ignore the details of variable formats. Numeric values are written as plain decimal numbers with enough digits to indicate their exact values in machine representation. Numeric values include 'e' followed by an exponent if the exponent value would be less than -4 or greater than 16. Dates are written in MM/DD/YYYY format and times in HH:MM:SS format. WKDAY and MONTH values are written as decimal numbers.

Numeric values use, by default, the decimal point character set with SET DECIMAL (see [SET DECIMAL], page 159). Use DECIMAL=DOT or DEC-IMAL=COMMA to force a particular decimal point character.

CELLS=VALUES FORMAT=VARIABLE

Writes variables using their print formats. Leading and trailing spaces are removed from numeric values, and trailing spaces are removed from string values.

CELLS=LABEL FORMAT=PLAIN
CELLS=LABEL FORMAT=VARIABLE

Writes value labels where they exist, and otherwise writes the values themselves as described above.

Regardless of CELLS and TEXTOPTIONS FORMAT, numeric system-missing values are output as a single space.

For TYPE=TAB, tab characters delimit values. For TYPE=CSV, the TEXTOPTIONS
DELIMITER and DECIMAL settings determine the character that separate values within a
line. If DELIMITER is specified, then the specified string separate values. If DELIMITER
is not specified, then the default is a comma with DECIMAL=DOT or a semicolon with
DECIMAL=COMMA. If DECIMAL is not given either, it is implied by the decimal point
character set with SET DECIMAL (see [SET DECIMAL], page 159).

The TEXTOPTIONS QUALIFIER setting specifies a character that is output before
and after a value that contains the delimiter character or the qualifier character. The
default is a double quote ('"'). A qualifier character that appears within a value is doubled.

9.8 SYSFILE INFO

SYSFILE INFO FILE='file_name' [ENCODING='encoding'].

SYSFILE INFO reads the dictionary in an SPSS system file, SPSS/PC+ system file, or
SPSS portable file, and displays the information in its dictionary.

Specify a file name or file handle. SYSFILE INFO reads that file and displays information
on its dictionary.

PSPP automatically detects the encoding of string data in the file, when possible.
The character encoding of old SPSS system files cannot always be guessed correctly,
and SPSS/PC+ system files do not include any indication of their encoding. Specify
the ENCODING subcommand with an IANA character set name as its string argument to
override the default, or specify ENCODING='DETECT' to analyze and report possibly valid
encodings for the system file. The ENCODING subcommand is a PSPP extension.

SYSFILE INFO does not affect the current active dataset.

9.9 XEXPORT

XEXPORT
    /OUTFILE='file_name'
    /DIGITS=n
    /DROP=var_list
    /KEEP=var_list
    /RENAME=(src_names=target_names)...
    /TYPE={COMM,TAPE}
    /MAP

The EXPORT transformation writes the active dataset dictionary and data to a specified
portable file.

This transformation is a PSPP extension.

It is similar to the EXPORT procedure, with two differences:

- XEXPORT is a transformation, not a procedure. It is executed when the data is read by
  a procedure or procedure-like command.

- XEXPORT does not support the UNSELECTED subcommand.

See Section 9.2 [EXPORT], page 82, for more information.

9.10 XSAVE

XSAVE
    /OUTFILE='*file_name*'
    /{UNCOMPRESSED,COMPRESSED,ZCOMPRESSED}
    /PERMISSIONS={WRITEABLE,READONLY}
    /DROP=*var_list*
    /KEEP=*var_list*
    /VERSION=*version*
    /RENAME=(*src_names=target_names*)...
    /NAMES
    /MAP

The XSAVE transformation writes the active dataset's dictionary and data to a system file. It is similar to the SAVE procedure, with two differences:

- XSAVE is a transformation, not a procedure. It is executed when the data is read by a procedure or procedure-like command.

- XSAVE does not support the UNSELECTED subcommand.

See Section 9.6 [SAVE], page 89, for more information.

10 Combining Data Files

This chapter describes commands that allow data from system files, portable files, and open datasets to be combined to form a new active dataset. These commands can combine data files in the following ways:

- ADD FILES interleaves or appends the cases from each input file. It is used with input files that have variables in common, but distinct sets of cases.

- MATCH FILES adds the data together in cases that match across multiple input files. It is used with input files that have cases in common, but different information about each case.

- UPDATE updates a master data file from data in a set of transaction files. Each case in a transaction data file modifies a matching case in the primary data file, or it adds a new case if no matching case can be found.

These commands share the majority of their syntax, which is described in the following section, followed by one section for each command that describes its specific syntax and semantics.

10.1 Common Syntax

```
Per input file:
    /FILE={*,'file_name'}
    [/RENAME=(src_names=target_names)...]
    [/IN=var_name]
    [/SORT]

Once per command:
    /BY var_list[({D|A})] [var_list[({D|A}]]...
    [/DROP=var_list]
    [/KEEP=var_list]
    [/FIRST=var_name]
    [/LAST=var_name]
    [/MAP]
```

This section describes the syntactical features in common among the ADD FILES, MATCH FILES, and UPDATE commands. The following sections describe details specific to each command.

Each of these commands reads two or more input files and combines them. The command's output becomes the new active dataset. None of the commands actually change the input files. Therefore, if you want the changes to become permanent, you must explicitly save them using an appropriate procedure or transformation (see Chapter 9 [System and Portable File IO], page 81).

The syntax of each command begins with a specification of the files to be read as input. For each input file, specify FILE with a system file or portable file's name as a string, a dataset (see Section 6.7 [Datasets], page 32) or file handle name, (see Section 6.9 [File Handles], page 44), or an asterisk ('*') to use the active dataset as input. Use of portable files on FILE is a PSPP extension.

At least two `FILE` subcommands must be specified. If the active dataset is used as an input source, then `TEMPORARY` must not be in effect.

Each `FILE` subcommand may be followed by any number of `RENAME` subcommands that specify a parenthesized group or groups of variable names as they appear in the input file, followed by those variables' new names, separated by an equals sign (=), e.g. `/RENAME=(OLD1=NEW1)(OLD2=NEW2)`. To rename a single variable, the parentheses may be omitted: `/RENAME=old=new`. Within a parenthesized group, variables are renamed simultaneously, so that `/RENAME=(A B=B A)` exchanges the names of variables *A* and *B*. Otherwise, renaming occurs in left-to-right order.

Each `FILE` subcommand may optionally be followed by a single `IN` subcommand, which creates a numeric variable with the specified name and format F1.0. The IN variable takes value 1 in an output case if the given input file contributed to that output case, and 0 otherwise. The `DROP`, `KEEP`, and `RENAME` subcommands have no effect on IN variables.

If `BY` is used (see below), the `SORT` keyword must be specified after a `FILE` if that input file is not already sorted on the `BY` variables. When `SORT` is specified, PSPP sorts the input file's data on the `BY` variables before it applies it to the command. When `SORT` is used, `BY` is required. `SORT` is a PSPP extension.

PSPP merges the dictionaries of all of the input files to form the dictionary of the new active dataset, like so:

- The variables in the new active dataset are the union of all the input files' variables, matched based on their name. When a single input file contains a variable with a given name, the output file will contain exactly that variable. When more than one input file contains a variable with a given name, those variables must all have the same type (numeric or string) and, for string variables, the same width. Variables are matched after renaming with the `RENAME` subcommand. Thus, `RENAME` can be used to resolve conflicts.

- The variable label for each output variable is taken from the first specified input file that has a variable label for that variable, and similarly for value labels and missing values.

- The file label of the new active dataset (see Section 16.12 [FILE LABEL], page 155) is that of the first specified `FILE` that has a file label.

- The documents in the new active dataset (see Section 16.5 [DOCUMENT], page 153) are the concatenation of all the input files' documents, in the order in which the `FILE` subcommands are specified.

- If all of the input files are weighted on the same variable, then the new active dataset is weighted on that variable. Otherwise, the new active dataset is not weighted.

The remaining subcommands apply to the output file as a whole, rather than to individual input files. They must be specified at the end of the command specification, following all of the `FILE` and related subcommands. The most important of these subcommands is `BY`, which specifies a set of one or more variables that may be used to find corresponding cases in each of the input files. The variables specified on `BY` must be present in all of the input files. Furthermore, if any of the input files are not sorted on the `BY` variables, then `SORT` must be specified for those input files.

The variables listed on BY may include (A) or (D) annotations to specify ascending or descending sort order. See Section 12.8 [SORT CASES], page 119, for more details on this notation. Adding (A) or (D) to the BY subcommand specification is a PSPP extension.

The DROP subcommand can be used to specify a list of variables to exclude from the output. By contrast, the KEEP subcommand can be used to specify variables to include in the output; all variables not listed are dropped. DROP and KEEP are executed in left-to-right order and may be repeated any number of times. DROP and KEEP do not affect variables created by the IN, FIRST, and LAST subcommands, which are always included in the new active dataset, but they can be used to drop BY variables.

The FIRST and LAST subcommands are optional. They may only be specified on MATCH FILES and ADD FILES, and only when BY is used. FIRST and LIST each adds a numeric variable to the new active dataset, with the name given as the subcommand's argument and F1.0 print and write formats. The value of the FIRST variable is 1 in the first output case with a given set of values for the BY variables, and 0 in other cases. Similarly, the LAST variable is 1 in the last case with a given of BY values, and 0 in other cases.

When any of these commands creates an output case, variables that are only in files that are not present for the current case are set to the system-missing value for numeric variables or spaces for string variables.

These commands may combine any number of files, limited only by the machine's memory.

10.2 ADD FILES

ADD FILES

Per input file:
    /FILE={*,'file_name'}
    [/RENAME=(src_names=target_names)...]
    [/IN=var_name]
    [/SORT]

Once per command:
    [/BY var_list[({D|A})] [var_list[({D|A})]...]]
    [/DROP=var_list]
    [/KEEP=var_list]
    [/FIRST=var_name]
    [/LAST=var_name]
    [/MAP]

ADD FILES adds cases from multiple input files. The output, which replaces the active dataset, consists all of the cases in all of the input files.

ADD FILES shares the bulk of its syntax with other PSPP commands for combining multiple data files. See Section 10.1 [Combining Files Common Syntax], page 95, above, for an explanation of this common syntax.

When BY is not used, the output of ADD FILES consists of all the cases from the first input file specified, followed by all the cases from the second file specified, and so on. When BY is used, the output is additionally sorted on the BY variables.

When `ADD FILES` creates an output case, variables that are not part of the input file from which the case was drawn are set to the system-missing value for numeric variables or spaces for string variables.

10.3 MATCH FILES

MATCH FILES

Per input file:
    /{FILE,TABLE}={*,'file_name'}
    [/RENAME=(src_names=target_names)...]
    [/IN=var_name]
    [/SORT]

Once per command:
    /BY var_list[({D|A}] [var_list[({D|A})]...]
    [/DROP=var_list]
    [/KEEP=var_list]
    [/FIRST=var_name]
    [/LAST=var_name]
    [/MAP]

`MATCH FILES` merges sets of corresponding cases in multiple input files into single cases in the output, combining their data.

`MATCH FILES` shares the bulk of its syntax with other PSPP commands for combining multiple data files. See Section 10.1 [Combining Files Common Syntax], page 95, above, for an explanation of this common syntax.

How `MATCH FILES` matches up cases from the input files depends on whether `BY` is specified:

- If `BY` is not used, `MATCH FILES` combines the first case from each input file to produce the first output case, then the second case from each input file for the second output case, and so on. If some input files have fewer cases than others, then the shorter files do not contribute to cases output after their input has been exhausted.

- If `BY` is used, `MATCH FILES` combines cases from each input file that have identical values for the `BY` variables.

When `BY` is used, `TABLE` subcommands may be used to introduce *table lookup file*. `TABLE` has same syntax as `FILE`, and the `RENAME`, `IN`, and `SORT` subcommands may follow a `TABLE` in the same way as `FILE`. Regardless of the number of `TABLE`s, at least one `FILE` must specified. Table lookup files are treated in the same way as other input files for most purposes and, in particular, table lookup files must be sorted on the `BY` variables or the `SORT` subcommand must be specified for that `TABLE`.

Cases in table lookup files are not consumed after they have been used once. This means that data in table lookup files can correspond to any number of cases in `FILE` input files. Table lookup files are analogous to lookup tables in traditional relational database systems.

If a table lookup file contains more than one case with a given set of `BY` variables, only the first case is used.

When `MATCH FILES` creates an output case, variables that are only in files that are not present for the current case are set to the system-missing value for numeric variables or spaces for string variables.

10.4 UPDATE

UPDATE

Per input file:
> /FILE={*,'file_name'}
> [/RENAME=(src_names=target_names)...]
> [/IN=var_name]
> [/SORT]

Once per command:
> /BY var_list[({D|A})] [var_list[({D|A})]]...
> [/DROP=var_list]
> [/KEEP=var_list]
> [/MAP]

UPDATE updates a *master file* by applying modifications from one or more *transaction files*.

UPDATE shares the bulk of its syntax with other PSPP commands for combining multiple data files. See Section 10.1 [Combining Files Common Syntax], page 95, above, for an explanation of this common syntax.

At least two **FILE** subcommands must be specified. The first **FILE** subcommand names the master file, and the rest name transaction files. Every input file must either be sorted on the variables named on the **BY** subcommand, or the **SORT** subcommand must be used just after the **FILE** subcommand for that input file.

UPDATE uses the variables specified on the **BY** subcommand, which is required, to attempt to match each case in a transaction file with a case in the master file:

- When a match is found, then the values of the variables present in the transaction file replace those variables' values in the new active file. If there are matching cases in more than more transaction file, PSPP applies the replacements from the first transaction file, then from the second transaction file, and so on. Similarly, if a single transaction file has cases with duplicate **BY** values, then those are applied in order to the master file.

  When a variable in a transaction file has a missing value or when a string variable's value is all blanks, that value is never used to update the master file.

- If a case in the master file has no matching case in any transaction file, then it is copied unchanged to the output.

- If a case in a transaction file has no matching case in the master file, then it causes a new case to be added to the output, initialized from the values in the transaction file.

11 Manipulating variables

The variables in the active dataset dictionary are important. There are several utility functions for examining and adjusting them.

11.1 ADD VALUE LABELS

> ADD VALUE LABELS
>   /*var_list* value 'label' [value 'label']...

ADD VALUE LABELS has the same syntax and purpose as VALUE LABELS (see Section 11.12 [VALUE LABELS], page 105), but it does not clear value labels from the variables before adding the ones specified.

11.2 DELETE VARIABLES

> DELETE VARIABLES *var_list*.

DELETE VARIABLES deletes the specified variables from the dictionary. It may not be used to delete all variables from the dictionary; use NEW FILE to do that (see Section 8.11 [NEW FILE], page 76).

DELETE VARIABLES should not be used after defining transformations but before executing a procedure. If it is used in such a context, it causes the data to be read. If it is used while TEMPORARY is in effect, it causes the temporary transformations to become permanent.

11.3 DISPLAY

> DISPLAY [SORTED] NAMES [[/VARIABLES=]*var_list*].
> DISPLAY [SORTED] INDEX [[/VARIABLES=]*var_list*].
> DISPLAY [SORTED] LABELS [[/VARIABLES=]*var_list*].
> DISPLAY [SORTED] VARIABLES [[/VARIABLES=]*var_list*].
> DISPLAY [SORTED] DICTIONARY [[/VARIABLES=]*var_list*].
> DISPLAY [SORTED] SCRATCH [[/VARIABLES=]*var_list*].
> DISPLAY [SORTED] ATTRIBUTES [[/VARIABLES=]*var_list*].
> DISPLAY [SORTED] @ATTRIBUTES [[/VARIABLES=]*var_list*].
> DISPLAY [SORTED] VECTORS.

DISPLAY displays information about the active dataset. A variety of different forms of information can be requested.

The following keywords primarily cause information about variables to be displayed. With these keywords, by default information is displayed about all variable in the active dataset, in the order that variables occur in the active dataset dictionary. The SORTED keyword causes output to be sorted alphabetically by variable name. The VARIABLES subcommand limits output to the specified variables.

NAMES   The variables' names are displayed.

INDEX   The variables' names are displayed along with a value describing their position within the active dataset dictionary.

LABELS   Variable names, positions, and variable labels are displayed.

VARIABLES

> Variable names, positions, print and write formats, and missing values are displayed.

DICTIONARY

> Variable names, positions, print and write formats, missing values, variable labels, and value labels are displayed.

SCRATCH

> Variable names are displayed, for scratch variables only (see Section 6.7.5 [Scratch Variables], page 43).

ATTRIBUTES
@ATTRIBUTES

> Datafile and variable attributes are displayed. The first form of the command omits those attributes whose names begin with @ or $@. In the second for, all datafile and variable attributes are displayed.

With the **VECTOR** keyword, **DISPLAY** lists all the currently declared vectors. If the **SORTED** keyword is given, the vectors are listed in alphabetical order; otherwise, they are listed in textual order of definition within the PSPP syntax file.

For related commands, see Section 16.6 [DISPLAY DOCUMENTS], page 154 and Section 16.7 [DISPLAY FILE LABEL], page 154.

11.4 FORMATS

> FORMATS var_list (fmt_spec) [var_list (fmt_spec)]. . . .

FORMATS set both print and write formats for the specified variables to the specified format specification. See Section 6.7.4 [Input and Output Formats], page 34.

Specify a list of variables followed by a format specification in parentheses. The print and write formats of the specified variables will be changed. All of the variables listed together must have the same type and, for string variables, the same width.

Additional lists of variables and formats may be included following the first one.

FORMATS takes effect immediately. It is not affected by conditional and looping structures such as DO IF or LOOP.

11.5 LEAVE

> LEAVE var_list.

LEAVE prevents the specified variables from being reinitialized whenever a new case is processed.

Normally, when a data file is processed, every variable in the active dataset is initialized to the system-missing value or spaces at the beginning of processing for each case. When a variable has been specified on **LEAVE**, this is not the case. Instead, that variable is initialized to 0 (not system-missing) or spaces for the first case. After that, it retains its value between cases.

This becomes useful for counters. For instance, in the example below the variable **SUM** maintains a running total of the values in the **ITEM** variable.

```
DATA LIST /ITEM 1-3.
COMPUTE SUM=SUM+ITEM.
PRINT /ITEM SUM.
LEAVE SUM
BEGIN DATA.
123
404
555
999
END DATA.
```

Partial output from this example:

```
123    123.00
404    527.00
555   1082.00
999   2081.00
```

It is best to use **LEAVE** command immediately before invoking a procedure command, because the left status of variables is reset by certain transformations—for instance, **COMPUTE** and **IF**. Left status is also reset by all procedure invocations.

11.6 MISSING VALUES

MISSING VALUES var_list (missing_values).

where *missing_values* takes one of the following forms:
> *num1*
> *num1, num2*
> *num1, num2, num3*
> *num1* THRU *num2*
> *num1* THRU *num2, num3*
> *string1*
> *string1, string2*
> *string1, string2, string3*

As part of a range, LO
or LOWEST
may take the place of *num1*;
HI
or HIGHEST
may take the place of *num2*.

MISSING VALUES sets user-missing values for numeric and string variables. Long string variables may have missing values, but characters after the first 8 bytes of the missing value must be spaces.

Specify a list of variables, followed by a list of their user-missing values in parentheses. Up to three discrete values may be given, or, for numeric variables only, a range of values optionally accompanied by a single discrete value. Ranges may be open-ended on one end, indicated through the use of the keyword **LO** or **LOWEST** or **HI** or **HIGHEST**.

The MISSING VALUES command takes effect immediately. It is not affected by conditional and looping constructs such as DO IF or LOOP.

11.7 MODIFY VARS

```
MODIFY VARS
    /REORDER={FORWARD,BACKWARD} {POSITIONAL,ALPHA} (var_list)...
    /RENAME=(old_names=new_names)...
    /{DROP,KEEP}=var_list
    /MAP
```

MODIFY VARS reorders, renames, and deletes variables in the active dataset.

At least one subcommand must be specified, and no subcommand may be specified more than once. DROP and KEEP may not both be specified.

The REORDER subcommand changes the order of variables in the active dataset. Specify one or more lists of variable names in parentheses. By default, each list of variables is rearranged into the specified order. To put the variables into the reverse of the specified order, put keyword BACKWARD before the parentheses. To put them into alphabetical order in the dictionary, specify keyword ALPHA before the parentheses. BACKWARD and ALPHA may also be combined.

To rename variables in the active dataset, specify RENAME, an equals sign ('='), and lists of the old variable names and new variable names separated by another equals sign within parentheses. There must be the same number of old and new variable names. Each old variable is renamed to the corresponding new variable name. Multiple parenthesized groups of variables may be specified.

The DROP subcommand deletes a specified list of variables from the active dataset.

The KEEP subcommand keeps the specified list of variables in the active dataset. Any unlisted variables are deleted from the active dataset.

MAP is currently ignored.

If either DROP or KEEP is specified, the data is read; otherwise it is not.

MODIFY VARS may not be specified following TEMPORARY (see Section 13.6 [TEMPORARY], page 122).

11.8 MRSETS

```
MRSETS
    /MDGROUP NAME=name VARIABLES=var_list VALUE=value
        [CATEGORYLABELS={VARLABELS,COUNTEDVALUES}]
        [{LABEL='label',LABELSOURCE=VARLABEL}]

    /MCGROUP NAME=name VARIABLES=var_list [LABEL='label']

    /DELETE NAME={[names],ALL}

    /DISPLAY NAME={[names],ALL}
```

MRSETS creates, modifies, deletes, and displays multiple response sets. A multiple response set is a set of variables that represent multiple responses to a single survey question in one of the two following ways:

- A *multiple dichotomy set* is analogous to a survey question with a set of checkboxes. Each variable in the set is treated in a Boolean fashion: one value (the "counted value") means that the box was checked, and any other value means that it was not.

- A *multiple category set* represents a survey question where the respondent is instructed to list up to *n* choices. Each variable represents one of the responses.

Any number of subcommands may be specified in any order.

The MDGROUP subcommand creates a new multiple dichotomy set or replaces an existing multiple response set. The NAME, VARIABLES, and VALUE specifications are required. The others are optional:

- *NAME* specifies the name used in syntax for the new multiple dichotomy set. The name must begin with '$'; it must otherwise follow the rules for identifiers (see Section 6.1 [Tokens], page 28).

- VARIABLES specifies the variables that belong to the set. At least two variables must be specified. The variables must be all string or all numeric.

- VALUE specifies the counted value. If the variables are numeric, the value must be an integer. If the variables are strings, then the value must be a string that is no longer than the shortest of the variables in the set (ignoring trailing spaces).

- CATEGORYLABELS optionally specifies the source of the labels for each category in the set:

  - VARLABELS, the default, uses variable labels or, for variables without variable labels, variable names. PSPP warns if two variables have the same variable label, since these categories cannot be distinguished in output.

  - COUNTEDVALUES instead uses each variable's value label for the counted value. PSPP warns if two variables have the same value label for the counted value or if one of the variables lacks a value label, since such categories cannot be distinguished in output.

- LABEL optionally specifies a label for the multiple response set. If neither LABEL nor LABELSOURCE=VARLABEL is specified, the set is unlabeled.

- LABELSOURCE=VARLABEL draws the multiple response set's label from the first variable label among the variables in the set; if none of the variables has a label, the name of the first variable is used. LABELSOURCE=VARLABEL must be used with CATEGORYLABELS=COUNTEDVALUES. It is mutually exclusive with LABEL.

The MCGROUP subcommand creates a new multiple category set or replaces an existing multiple response set. The NAME and VARIABLES specifications are required, and LABEL is optional. Their meanings are as described above in MDGROUP. PSPP warns if two variables in the set have different value labels for a single value, since each of the variables in the set should have the same possible categories.

The DELETE subcommand deletes multiple response groups. A list of groups may be named within a set of required square brackets, or ALL may be used to delete all groups.

The DISPLAY subcommand displays information about defined multiple response sets. Its syntax is the same as the DELETE subcommand.

Multiple response sets are saved to and read from system files by, e.g., the **SAVE** and **GET** command. Otherwise, multiple response sets are currently used only by third party software.

11.9 NUMERIC

NUMERIC /*var_list* [(*fmt_spec*)].

NUMERIC explicitly declares new numeric variables, optionally setting their output formats.

Specify a slash ('/'), followed by the names of the new numeric variables. If you wish to set their output formats, follow their names by an output format specification in parentheses (see Section 6.7.4 [Input and Output Formats], page 34); otherwise, the default is F8.2.

Variables created with **NUMERIC** are initialized to the system-missing value.

11.10 PRINT FORMATS

PRINT FORMATS *var_list* (*fmt_spec*) [*var_list* (*fmt_spec*)]. . . .

PRINT FORMATS sets the print formats for the specified variables to the specified format specification.

Its syntax is identical to that of **FORMATS** (see Section 11.4 [FORMATS], page 101), but **PRINT FORMATS** sets only print formats, not write formats.

11.11 RENAME VARIABLES

RENAME VARIABLES (*old_names=new_names*). . . .

RENAME VARIABLES changes the names of variables in the active dataset. Specify lists of the old variable names and new variable names, separated by an equals sign ('='), within parentheses. There must be the same number of old and new variable names. Each old variable is renamed to the corresponding new variable name. Multiple parenthesized groups of variables may be specified. When the old and new variable names contain only a single variable name, the parentheses are optional.

RENAME VARIABLES takes effect immediately. It does not cause the data to be read.

RENAME VARIABLES may not be specified following **TEMPORARY** (see Section 13.6 [TEMPORARY], page 122).

11.12 VALUE LABELS

VALUE LABELS
    /*var_list value* 'label' [*value* 'label']. . .

VALUE LABELS allows values of numeric and short string variables to be associated with labels. In this way, a short value can stand for a long value.

To set up value labels for a set of variables, specify the variable names after a slash ('/'), followed by a list of values and their associated labels, separated by spaces.

Value labels in output are normally broken into lines automatically. Put '\n' in a label string to force a line break at that point. The label may still be broken into lines at additional points.

Before **VALUE LABELS** is executed, any existing value labels are cleared from the variables specified. Use **ADD VALUE LABELS** (see Section 11.1 [ADD VALUE LABELS], page 100) to add value labels without clearing those already present.

11.13 STRING

STRING *var_list* (*fmt_spec*) [/ *var_list* (*fmt_spec*)] [...].

STRING creates new string variables for use in transformations.

Specify a list of names for the variable you want to create, followed by the desired output format specification in parentheses (see Section 6.7.4 [Input and Output Formats], page 34). Variable widths are implicitly derived from the specified output formats. The created variables will be initialized to spaces.

If you want to create several variables with distinct output formats, you can either use two or more separate **STRING** commands, or you can specify further variable list and format specification pairs, each separated from the previous by a slash ('/').

The following example is one way to create three string variables; Two of the variables have format A24 and the other A80:

```
STRING firstname lastname (A24) / address (A80).
```

Here is another way to achieve the same result:

```
STRING firstname lastname (A24).
STRING address (A80).
```

... and here is yet another way:

```
STRING firstname (A24).
STRING lastname (A24).
STRING address (A80).
```

11.14 VARIABLE ATTRIBUTE

VARIABLE ATTRIBUTE
    VARIABLES=*var_list*
    ATTRIBUTE=*name*('value') [*name*('value')]...
    ATTRIBUTE=*name*[*index*]('value') [*name*[*index*]('value')]...
    DELETE=*name* [*name*]...
    DELETE=*name*[*index*] [*name*[*index*]]...

VARIABLE ATTRIBUTE adds, modifies, or removes user-defined attributes associated with variables in the active dataset. Custom variable attributes are not interpreted by PSPP, but they are saved as part of system files and may be used by other software that reads them.

The required **VARIABLES** subcommand must come first. Specify the variables to which the following **ATTRIBUTE** or **DELETE** subcommand should apply.

Use the **ATTRIBUTE** subcommand to add or modify custom variable attributes. Specify the name of the attribute as an identifier (see Section 6.1 [Tokens], page 28), followed by the desired value, in parentheses, as a quoted string. The specified attributes are then added or modified in the variables specified on **VARIABLES**. Attribute names that begin with $ are reserved for PSPP's internal use, and attribute names that begin with @ or $@ are not displayed by most PSPP commands that display other attributes. Other attribute names are not treated specially.

Attributes may also be organized into arrays. To assign to an array element, add an integer array index enclosed in square brackets ([and]) between the attribute name and value. Array indexes start at 1, not 0. An attribute array that has a single element (number 1) is not distinguished from a non-array attribute.

Use the DELETE subcommand to delete an attribute from the variable specified on VARIABLES. Specify an attribute name by itself to delete an entire attribute, including all array elements for attribute arrays. Specify an attribute name followed by an array index in square brackets to delete a single element of an attribute array. In the latter case, all the array elements numbered higher than the deleted element are shifted down, filling the vacated position.

To associate custom attributes with the entire active dataset, instead of with particular variables, use DATAFILE ATTRIBUTE (see Section 8.3 [DATAFILE ATTRIBUTE], page 64) instead.

VARIABLE ATTRIBUTE takes effect immediately. It is not affected by conditional and looping structures such as DO IF or LOOP.

11.15 VARIABLE LABELS

VARIABLE LABELS
        var_list 'var_label'
        [/var_list 'var_label']
        .
        .
        .
        [/var_list 'var_label']

VARIABLE LABELS associates explanatory names with variables. This name, called a *variable label*, is displayed by statistical procedures.

To assign a variable label to a group of variables, specify a list of variable names and the variable label as a string. To assign different labels to different variables in the same command, precede the subsequent variable list with a slash ('/').

11.16 VARIABLE ALIGNMENT

VARIABLE ALIGNMENT
        var_list (LEFT | RIGHT | CENTER)
        [/var_list (LEFT | RIGHT | CENTER)]
        .
        .
        .
        [/var_list (LEFT | RIGHT | CENTER)]

VARIABLE ALIGNMENT sets the alignment of variables for display editing purposes. This only has effect for third party software. It does not affect the display of variables in the PSPP output.

11.17 VARIABLE WIDTH

VARIABLE WIDTH
    var_list (width)
    [/ *var_list* (width)]
       .
       .
       .
    [/ *var_list* (width)]

VARIABLE WIDTH sets the column width of variables for display editing purposes. This only affects third party software. It does not affect the display of variables in the PSPP output.

11.18 VARIABLE LEVEL

VARIABLE LEVEL
    var_list (SCALE | NOMINAL | ORDINAL)
    [/ *var_list* (SCALE | NOMINAL | ORDINAL)]
       .
       .
       .
    [/ *var_list* (SCALE | NOMINAL | ORDINAL)]

VARIABLE LEVEL sets the measurement level of variables. Currently, this has no effect except for certain third party software.

11.19 VARIABLE ROLE

VARIABLE ROLE
    /*role var_list*
    [/*role var_list*]...

VARIABLE ROLE sets the intended role of a variable for use in dialog boxes in graphical user interfaces. Each *role* specifies one of the following roles for the variables that follow it:

INPUT   An input variable, such as an independent variable.

TARGET   An output variable, such as an dependent variable.

BOTH   A variable used for input and output.

NONE   No role assigned. (This is a variable's default role.)

PARTITION
    Used to break the data into groups for testing.

SPLIT   No meaning except for certain third party software. (This role's meaning is unrelated to SPLIT FILE.)

The PSPPIRE GUI does not yet use variable roles as intended.

11.20 VECTOR

Two possible syntaxes:

VECTOR *vec_name=var_list*.

VECTOR *vec_name_list(count [format])*.

VECTOR allows a group of variables to be accessed as if they were consecutive members of an array with a vector(index) notation.

To make a vector out of a set of existing variables, specify a name for the vector followed by an equals sign ('=') and the variables to put in the vector. All the variables in the vector must be the same type. String variables in a vector must all have the same width.

To make a vector and create variables at the same time, specify one or more vector names followed by a count in parentheses. This will cause variables named *vec1* through *veccount* to be created as numeric variables. By default, the new variables have print and write format F8.2, but an alternate format may be specified inside the parentheses before or after the count and separated from it by white space or a comma. Variable names including numeric suffixes may not exceed 64 characters in length, and none of the variables may exist prior to VECTOR.

Vectors created with VECTOR disappear after any procedure or procedure-like command is executed. The variables contained in the vectors remain, unless they are scratch variables (see Section 6.7.5 [Scratch Variables], page 43).

Variables within a vector may be referenced in expressions using vector(index) syntax.

11.21 WRITE FORMATS

WRITE FORMATS *var_list (fmt_spec)* [*var_list (fmt_spec)*]. ...

WRITE FORMATS sets the write formats for the specified variables to the specified format specification. Its syntax is identical to that of FORMATS (see Section 11.4 [FORMATS], page 101), but WRITE FORMATS sets only write formats, not print formats.

12 Data transformations

The PSPP procedures examined in this chapter manipulate data and prepare the active dataset for later analyses. They do not produce output, as a rule.

12.1 AGGREGATE

```
AGGREGATE
    OUTFILE={*,'file_name',file_handle} [MODE={REPLACE, ADDVARIABLES}]
      /PRESORTED
      /DOCUMENT
      /MISSING=COLUMNWISE
      /BREAK=var_list
      /dest_var['label']...=agr_func(src_vars, args...)...
```

AGGREGATE summarizes groups of cases into single cases. Cases are divided into groups that have the same values for one or more variables called *break variables*. Several functions are available for summarizing case contents.

The OUTFILE subcommand is required and must appear first. Specify a system file or portable file by file name or file handle (see Section 6.9 [File Handles], page 44), or a dataset by its name (see Section 6.7 [Datasets], page 32). The aggregated cases are written to this file. If '*' is specified, then the aggregated cases replace the active dataset's data. Use of OUTFILE to write a portable file is a PSPP extension.

If OUTFILE=* is given, then the subcommand MODE may also be specified. The mode subcommand has two possible values: ADDVARIABLES or REPLACE. In REPLACE mode, the entire active dataset is replaced by a new dataset which contains just the break variables and the destination varibles. In this mode, the new file will contain as many cases as there are unique combinations of the break variables. In ADDVARIABLES mode, the destination variables will be appended to the existing active dataset. Cases which have identical combinations of values in their break variables, will receive identical values for the destination variables. The number of cases in the active dataset will remain unchanged. Note that if ADDVARIABLES is specified, then the data *must* be sorted on the break variables.

By default, the active dataset will be sorted based on the break variables before aggregation takes place. If the active dataset is already sorted or otherwise grouped in terms of the break variables, specify PRESORTED to save time. PRESORTED is assumed if MODE=ADDVARIABLES is used.

Specify DOCUMENT to copy the documents from the active dataset into the aggregate file (see Section 16.5 [DOCUMENT], page 153). Otherwise, the aggregate file will not contain any documents, even if the aggregate file replaces the active dataset.

Normally, only a single case (for SD and SD., two cases) need be non-missing in each group for the aggregate variable to be non-missing. Specifying /MISSING=COLUMNWISE inverts this behavior, so that the aggregate variable becomes missing if any aggregated value is missing.

If PRESORTED, DOCUMENT, or MISSING are specified, they must appear between OUTFILE and BREAK.

At least one break variable must be specified on BREAK, a required subcommand. The values of these variables are used to divide the active dataset into groups to be summarized. In addition, at least one *dest_var* must be specified.

One or more sets of aggregation variables must be specified. Each set comprises a list of aggregation variables, an equals sign ('='), the name of an aggregation function (see the list below), and a list of source variables in parentheses. Some aggregation functions expect additional arguments following the source variable names.

Aggregation variables typically are created with no variable label, value labels, or missing values. Their default print and write formats depend on the aggregation function used, with details given in the table below. A variable label for an aggregation variable may be specified just after the variable's name in the aggregation variable list.

Each set must have exactly as many source variables as aggregation variables. Each aggregation variable receives the results of applying the specified aggregation function to the corresponding source variable. The MEAN, MEDIAN, SD, and SUM aggregation functions may only be applied to numeric variables. All the rest may be applied to numeric and string variables.

The available aggregation functions are as follows:

FGT(var_name, value)

> Fraction of values greater than the specified constant. The default format is F5.3.

FIN(var_name, low, high)

> Fraction of values within the specified inclusive range of constants. The default format is F5.3.

FLT(var_name, value)

> Fraction of values less than the specified constant. The default format is F5.3.

FIRST(var_name)

> First non-missing value in break group. The aggregation variable receives the complete dictionary information from the source variable. The sort performed by AGGREGATE (and by SORT CASES) is stable, so that the first case with particular values for the break variables before sorting will also be the first case in that break group after sorting.

FOUT(var_name, low, high)

> Fraction of values strictly outside the specified range of constants. The default format is F5.3.

LAST(var_name)

> Last non-missing value in break group. The aggregation variable receives the complete dictionary information from the source variable. The sort performed by AGGREGATE (and by SORT CASES) is stable, so that the last case with particular values for the break variables before sorting will also be the last case in that break group after sorting.

MAX(var_name)

> Maximum value. The aggregation variable receives the complete dictionary information from the source variable.

MEAN(var_name)

> Arithmetic mean. Limited to numeric values. The default format is F8.2.

MEDIAN(*var_name*)

> The median value. Limited to numeric values. The default format is F8.2.

MIN(*var_name*)

> Minimum value. The aggregation variable receives the complete dictionary information from the source variable.

N(*var_name*)

> Number of non-missing values. The default format is F7.0 if weighting is not enabled, F8.2 if it is (see Section 13.7 [WEIGHT], page 123).

N

> Number of cases aggregated to form this group. The default format is F7.0 if weighting is not enabled, F8.2 if it is (see Section 13.7 [WEIGHT], page 123).

NMISS(*var_name*)

> Number of missing values. The default format is F7.0 if weighting is not enabled, F8.2 if it is (see Section 13.7 [WEIGHT], page 123).

NU(*var_name*)

> Number of non-missing values. Each case is considered to have a weight of 1, regardless of the current weighting variable (see Section 13.7 [WEIGHT], page 123). The default format is F7.0.

NU

> Number of cases aggregated to form this group. Each case is considered to have a weight of 1, regardless of the current weighting variable. The default format is F7.0.

NUMISS(*var_name*)

> Number of missing values. Each case is considered to have a weight of 1, regardless of the current weighting variable. The default format is F7.0.

PGT(*var_name*, *value*)

> Percentage between 0 and 100 of values greater than the specified constant. The default format is F5.1.

PIN(*var_name*, *low*, *high*)

> Percentage of values within the specified inclusive range of constants. The default format is F5.1.

PLT(*var_name*, *value*)

> Percentage of values less than the specified constant. The default format is F5.1.

POUT(*var_name*, *low*, *high*)

> Percentage of values strictly outside the specified range of constants. The default format is F5.1.

SD(*var_name*)

> Standard deviation of the mean. Limited to numeric values. The default format is F8.2.

SUM(*var_name*)

> Sum. Limited to numeric values. The default format is F8.2.

Aggregation functions compare string values in terms of internal character codes. On most modern computers, this is ASCII or a superset thereof.

The aggregation functions listed above exclude all user-missing values from calculations. To include user-missing values, insert a period ('.') at the end of the function name. (e.g. 'SUM.'). (Be aware that specifying such a function as the last token on a line will cause the period to be interpreted as the end of the command.)

AGGREGATE both ignores and cancels the current SPLIT FILE settings (see Section 13.5 [SPLIT FILE], page 121).

12.2 AUTORECODE

AUTORECODE VARIABLES=*src_vars* INTO *dest_vars*
    [/DESCENDING]
    [/PRINT]
    [/GROUP]
    [/BLANK = {VALID, MISSING}]

The AUTORECODE procedure considers the *n* values that a variable takes on and maps them onto values 1...*n* on a new numeric variable.

Subcommand VARIABLES is the only required subcommand and must come first. Specify VARIABLES, an equals sign ('='), a list of source variables, INTO, and a list of target variables. There must the same number of source and target variables. The target variables must not already exist.

By default, increasing values of a source variable (for a string, this is based on character code comparisons) are recoded to increasing values of its target variable. To cause increasing values of a source variable to be recoded to decreasing values of its target variable (*n* down to 1), specify DESCENDING.

PRINT is currently ignored.

The GROUP subcommand is relevant only if more than one variable is to be recoded. It causes a single mapping between source and target values to be used, instead of one map per variable.

If /BLANK=MISSING is given, then string variables which contain only whitespace are recoded as SYSMIS. If /BLANK=VALID is given then they will be allocated a value like any other. /BLANK is not relevant to numeric values. /BLANK=VALID is the default.

AUTORECODE is a procedure. It causes the data to be read.

12.3 COMPUTE

COMPUTE *variable* = *expression*.

or

COMPUTE vector(*index*) = *expression*.

COMPUTE assigns the value of an expression to a target variable. For each case, the expression is evaluated and its value assigned to the target variable. Numeric and string variables may be assigned. When a string expression's width differs from the target variable's width, the string result of the expression is truncated or padded with spaces on the right as necessary. The expression and variable types must match.

For numeric variables only, the target variable need not already exist. Numeric variables created by `COMPUTE` are assigned an `F8.2` output format. String variables must be declared before they can be used as targets for `COMPUTE`.

The target variable may be specified as an element of a vector (see Section 11.20 [VECTOR], page 109). In this case, an expression *index* must be specified in parentheses following the vector name. The expression *index* must evaluate to a numeric value that, after rounding down to the nearest integer, is a valid index for the named vector.

Using `COMPUTE` to assign to a variable specified on `LEAVE` (see Section 11.5 [LEAVE], page 101) resets the variable's left state. Therefore, `LEAVE` should be specified following `COMPUTE`, not before.

`COMPUTE` is a transformation. It does not cause the active dataset to be read.

When `COMPUTE` is specified following `TEMPORARY` (see Section 13.6 [TEMPORARY], page 122), the `LAG` function may not be used (see [LAG], page 57).

12.4 COUNT

COUNT *var_name* = *var*... (*value*...).

Each *value* takes one of the following forms:
　　　number
　　　string
　　　num1 THRU *num2*
　　　MISSING
　　　SYSMIS
where *num1* is a numeric expression or the words `LO`
　or `LOWEST`

　　　and *num2* is a numeric expression or `HI`
　or `HIGHEST`

　.

`COUNT` creates or replaces a numeric *target* variable that counts the occurrence of a *criterion* value or set of values over one or more *test* variables for each case.

The target variable values are always nonnegative integers. They are never missing. The target variable is assigned an F8.2 output format. See Section 6.7.4 [Input and Output Formats], page 34. Any variables, including string variables, may be test variables.

User-missing values of test variables are treated just like any other values. They are **not** treated as system-missing values. User-missing values that are criterion values or inside ranges of criterion values are counted as any other values. However (for numeric variables), keyword `MISSING` may be used to refer to all system- and user-missing values.

`COUNT` target variables are assigned values in the order specified. In the command `COUNT A=A B(1) /B=A B(2).`, the following actions occur:

− The number of occurrences of 1 between *A* and *B* is counted.

− *A* is assigned this value.

− The number of occurrences of 1 between *B* and the **new** value of *A* is counted.

− *B* is assigned this value.

Despite this ordering, all `COUNT` criterion variables must exist before the procedure is executed—they may not be created as target variables earlier in the command! Break such a command into two separate commands.

The examples below may help to clarify.

A. Assuming `Q0`, `Q2`, ..., `Q9` are numeric variables, the following commands:

1. Count the number of times the value 1 occurs through these variables for each case and assigns the count to variable `QCOUNT`.

2. Print out the total number of times the value 1 occurs throughout *all* cases using `DESCRIPTIVES`. See Section 15.1 [DESCRIPTIVES], page 127, for details.

```
COUNT QCOUNT=Q0 TO Q9(1).
DESCRIPTIVES QCOUNT /STATISTICS=SUM.
```

B. Given these same variables, the following commands:

1. Count the number of valid values of these variables for each case and assigns the count to variable `QVALID`.

2. Multiplies each value of `QVALID` by 10 to obtain a percentage of valid values, using `COMPUTE`. See Section 12.3 [COMPUTE], page 113, for details.

3. Print out the percentage of valid values across all cases, using `DESCRIPTIVES`. See Section 15.1 [DESCRIPTIVES], page 127, for details.

```
COUNT QVALID=Q0 TO Q9 (LO THRU HI).
COMPUTE QVALID=QVALID*10.
DESCRIPTIVES QVALID /STATISTICS=MEAN.
```

12.5 FLIP

> FLIP /VARIABLES=*var_list* /NEWNAMES=*var_name*.

`FLIP` transposes rows and columns in the active dataset. It causes cases to be swapped with variables, and vice versa.

All variables in the transposed active dataset are numeric. String variables take on the system-missing value in the transposed file.

N subcommands are required. If specified, the `VARIABLES` subcommand selects variables to be transformed into cases, and variables not specified are discarded. If the `VARIABLES` subcommand is omitted, all variables are selected for transposition.

The variables specified by `NEWNAMES`, which must be a string variable, is used to give names to the variables created by `FLIP`. Only the first 8 characters of the variable are used. If `NEWNAMES` is not specified then the default is a variable named CASE_LBL, if it exists. If it does not then the variables created by `FLIP` are named VAR000 through VAR999, then VAR1000, VAR1001, and so on.

When a `NEWNAMES` variable is available, the names must be canonicalized before becoming variable names. Invalid characters are replaced by letter 'V' in the first position, or by '_' in subsequent positions. If the name thus generated is not unique, then numeric extensions are added, starting with 1, until a unique name is found or there are no remaining possibilities. If the latter occurs then the `FLIP` operation aborts.

The resultant dictionary contains a CASE_LBL variable, a string variable of width 8, which stores the names of the variables in the dictionary before the transposition. Variables names longer than 8 characters are truncated. If the active dataset is subsequently transposed using FLIP, this variable can be used to recreate the original variable names.

FLIP honors N OF CASES (see Section 13.2 [N OF CASES], page 120). It ignores TEMPORARY (see Section 13.6 [TEMPORARY], page 122), so that "temporary" transformations become permanent.

12.6 IF

> IF *condition variable=expression.*

or

> IF *condition vector(index)=expression.*

The IF transformation conditionally assigns the value of a target expression to a target variable, based on the truth of a test expression.

Specify a boolean-valued expression (see Chapter 7 [Expressions], page 46) to be tested following the IF keyword. This expression is evaluated for each case. If the value is true, then the value of the expression is computed and assigned to the specified variable. If the value is false or missing, nothing is done. Numeric and string variables may be assigned. When a string expression's width differs from the target variable's width, the string result of the expression is truncated or padded with spaces on the right as necessary. The expression and variable types must match.

The target variable may be specified as an element of a vector (see Section 11.20 [VECTOR], page 109). In this case, a vector index expression must be specified in parentheses following the vector name. The index expression must evaluate to a numeric value that, after rounding down to the nearest integer, is a valid index for the named vector.

Using IF to assign to a variable specified on LEAVE (see Section 11.5 [LEAVE], page 101) resets the variable's left state. Therefore, LEAVE should be specified following IF, not before.

When IF is specified following TEMPORARY (see Section 13.6 [TEMPORARY], page 122), the LAG function may not be used (see [LAG], page 57).

12.7 RECODE

The RECODE command is used to transform existing values into other, user specified values. The general form is:

> RECODE *src_vars*
>   (*src_value src_value* ... = *dest_value*)
>   (*src_value src_value* ... = *dest_value*)
>   (*src_value src_value* ... = *dest_value*) ...
>   [INTO *dest_vars*].

Following the RECODE keyword itself comes *src_vars* which is a list of variables whose values are to be transformed. These variables may be string variables or they may be numeric. However the list must be homogeneous; you may not mix string variables and numeric variables in the same recoding.

After the list of source variables, there should be one or more *mappings*. Each mapping is enclosed in parentheses, and contains the source values and a destination value separated by a single '='. The source values are used to specify the values in the dataset which need to change, and the destination value specifies the new value to which they should be changed. Each *src_value* may take one of the following forms:

number   If the source variables are numeric then *src_value* may be a literal number.

string   If the source variables are string variables then *src_value* may be a literal string (like all strings, enclosed in single or double quotes).

num1 THRU *num2*

This form is valid only when the source variables are numeric. It specifies all values in the range between *num1* and *num2*, including both endpoints of the range. By convention, *num1* should be less than *num2*. Open-ended ranges may be specified using 'LO' or 'LOWEST' for *num1* or 'HI' or 'HIGHEST' for *num2*.

'MISSING'   The literal keyword 'MISSING' matches both system missing and user missing values. It is valid for both numeric and string variables.

'SYSMIS'   The literal keyword 'SYSMIS' matches system missing values. It is valid for both numeric variables only.

'ELSE'   The 'ELSE' keyword may be used to match any values which are not matched by any other *src_value* appearing in the command. If this keyword appears, it should be used in the last mapping of the command.

After the source variables comes an '=' and then the *dest_value*. The *dest_value* may take any of the following forms:

number   A literal numeric value to which the source values should be changed. This implies the destination variable must be numeric.

string   A literal string value (enclosed in quotation marks) to which the source values should be changed. This implies the destination variable must be a string variable.

'SYSMIS'   The keyword 'SYSMIS' changes the value to the system missing value. This implies the destination variable must be numeric.

'COPY'   The special keyword 'COPY' means that the source value should not be modified, but copied directly to the destination value. This is meaningful only if 'INTO dest_vars' is specified.

Mappings are considered from left to right. Therefore, if a value is matched by a *src_value* from more than one mapping, the first (leftmost) mapping which matches will be considered. Any subsequent matches will be ignored.

The clause 'INTO dest_vars' is optional. The behaviour of the command is slightly different depending on whether it appears or not.

If 'INTO dest_vars' does not appear, then values will be recoded "in place". This means that the recoded values are written back to the source variables from whence the original values came. In this case, the *dest_value* for every mapping must imply a value which has the same type as the *src_value*. For example, if the source value is a string value, it is not permissible for *dest_value* to be 'SYSMIS' or another forms which implies a numeric result. It is also not permissible for *dest_value* to be longer than the width of the source variable.

The following example two numeric variables *x* and *y* are recoded in place. Zero is recoded to 99, the values 1 to 10 inclusive are unchanged, values 1000 and higher are recoded to the system-missing value and all other values are changed to 999:

```
recode x y
        (0 = 99)
        (1 THRU 10 = COPY)
        (1000 THRU HIGHEST = SYSMIS)
        (ELSE = 999).
```

If 'INTO *dest_vars*' is given, then recoded values are written into the variables specified in *dest_vars*, which must therefore contain a list of valid variable names. The number of variables in *dest_vars* must be the same as the number of variables in *src_vars* and the respective order of the variables in *dest_vars* corresponds to the order of *src_vars*. That is to say, recoded values whose original value came from the *n*th variable in *src_vars* will be placed into the *n*th variable in *dest_vars*. The source variables will be unchanged. If any mapping implies a string as its destination value, then the respective destination variable must already exist, or have been declared using STRING or another transformation. Numeric variables however will be automatically created if they don't already exist. The following example deals with two source variables, *a* and *b* which contain string values. Hence there are two destination variables *v1* and *v2*. Any cases where *a* or *b* contain the values 'apple', 'pear' or 'pomegranate' will result in *v1* or *v2* being filled with the string 'fruit' whilst cases with 'tomato', 'lettuce' or 'carrot' will result in 'vegetable'. Any other values will produce the result 'unknown':

```
string v1 (a20).
string v2 (a20).

recode a b
        ("apple" "pear" "pomegranate" = "fruit")
        ("tomato" "lettuce" "carrot" = "vegetable")
        (ELSE = "unknown")
        into v1 v2.
```

There is one very special mapping, not mentioned above. If the source variable is a string variable then a mapping may be specified as '(CONVERT)'. This mapping, if it appears must be the last mapping given and the 'INTO *dest_vars*' clause must also be given and must not refer to a string variable. 'CONVERT' causes a number specified as a string to be converted to a numeric value. For example it will convert the string '"3"' into the numeric value 3 (note that it will not convert 'three' into 3). If the string cannot be parsed as a number, then the system-missing value is assigned instead. In the following example, cases where the value of *x* (a string variable) is the empty string, are recoded to 999 and all others are converted to the numeric equivalent of the input value. The results are placed into the numeric variable *y*:

```
recode x
        ("" = 999)
        (convert)
        into y.
```

It is possible to specify multiple recodings on a single command. Introduce additional recodings with a slash ('/') to separate them from the previous recodings:

```
recode
        a  (2 = 22) (else = 99)
        /b (1 = 3) into z
        .
```

Here we have two recodings. The first affects the source variable a and recodes in-place the value 2 into 22 and all other values to 99. The second recoding copies the values of b into the variable z, changing any instances of 1 into 3.

12.8 SORT CASES

SORT CASES BY *var_list*[({D | A}] [*var_list*[({D | A}]] ...

SORT CASES sorts the active dataset by the values of one or more variables.

Specify BY and a list of variables to sort by. By default, variables are sorted in ascending order. To override sort order, specify (D) or (DOWN) after a list of variables to get descending order, or (A) or (UP) for ascending order. These apply to all the listed variables up until the preceding (A), (D), (UP) or (DOWN).

The sort algorithms used by SORT CASES are stable. That is, records that have equal values of the sort variables will have the same relative order before and after sorting. As a special case, re-sorting an already sorted file will not affect the ordering of cases.

SORT CASES is a procedure. It causes the data to be read.

SORT CASES attempts to sort the entire active dataset in main memory. If workspace is exhausted, it falls back to a merge sort algorithm that involves creates numerous temporary files.

SORT CASES may not be specified following TEMPORARY.

13 Selecting data for analysis

This chapter documents PSPP commands that temporarily or permanently select data records from the active dataset for analysis.

13.1 FILTER

> FILTER BY var_name.
> FILTER OFF.

FILTER allows a boolean-valued variable to be used to select cases from the data stream for processing.

To set up filtering, specify BY and a variable name. Keyword BY is optional but recommended. Cases which have a zero or system- or user-missing value are excluded from analysis, but not deleted from the data stream. Cases with other values are analyzed. To filter based on a different condition, use transformations such as COMPUTE or RECODE to compute a filter variable of the required form, then specify that variable on FILTER.

FILTER OFF turns off case filtering.

Filtering takes place immediately before cases pass to a procedure for analysis. Only one filter variable may be active at a time. Normally, case filtering continues until it is explicitly turned off with FILTER OFF. However, if FILTER is placed after TEMPORARY, it filters only the next procedure or procedure-like command.

13.2 N OF CASES

> N [OF CASES] num_of_cases [ESTIMATED].

N OF CASES limits the number of cases processed by any procedures that follow it in the command stream. N OF CASES 100, for example, tells PSPP to disregard all cases after the first 100.

When N OF CASES is specified after TEMPORARY, it affects only the next procedure (see Section 13.6 [TEMPORARY], page 122). Otherwise, cases beyond the limit specified are not processed by any later procedure.

If the limit specified on N OF CASES is greater than the number of cases in the active dataset, it has no effect.

When N OF CASES is used along with SAMPLE or SELECT IF, the case limit is applied to the cases obtained after sampling or case selection, regardless of how N OF CASES is placed relative to SAMPLE or SELECT IF in the command file. Thus, the commands N OF CASES 100 and SAMPLE .5 will both randomly sample approximately half of the active dataset's cases, then select the first 100 of those sampled, regardless of their order in the command file.

N OF CASES with the ESTIMATED keyword gives an estimated number of cases before DATA LIST or another command to read in data. ESTIMATED never limits the number of cases processed by procedures. PSPP currently does not make use of case count estimates.

13.3 SAMPLE

SAMPLE *num1* [FROM *num2*].

SAMPLE randomly samples a proportion of the cases in the active file. Unless it follows **TEMPORARY**, it operates as a transformation, permanently removing cases from the active dataset.

The proportion to sample can be expressed as a single number between 0 and 1. If k is the number specified, and N is the number of currently-selected cases in the active dataset, then after **SAMPLE** k., approximately $k*N$ cases will be selected.

The proportion to sample can also be specified in the style **SAMPLE** m **FROM** N. With this style, cases are selected as follows:

1. If N is equal to the number of currently-selected cases in the active dataset, exactly m cases will be selected.

2. If N is greater than the number of currently-selected cases in the active dataset, an equivalent proportion of cases will be selected.

3. If N is less than the number of currently-selected cases in the active, exactly m cases will be selected *from the first N cases in the active dataset.*

SAMPLE and **SELECT IF** are performed in the order specified by the syntax file.

SAMPLE is always performed before **N OF CASES**, regardless of ordering in the syntax file (see Section 13.2 [N OF CASES], page 120).

The same values for **SAMPLE** may result in different samples. To obtain the same sample, use the **SET** command to set the random number seed to the same value before each **SAMPLE**. Different samples may still result when the file is processed on systems with differing endianness or floating-point formats. By default, the random number seed is based on the system time.

13.4 SELECT IF

SELECT IF *expression*.

SELECT IF selects cases for analysis based on the value of *expression*. Cases not selected are permanently eliminated from the active dataset, unless **TEMPORARY** is in effect (see Section 13.6 [TEMPORARY], page 122).

Specify a boolean expression (see Chapter 7 [Expressions], page 46). If the value of the expression is true for a particular case, the case will be analyzed. If the expression has a false or missing value, then the case will be deleted from the data stream.

Place **SELECT IF** as early in the command file as possible. Cases that are deleted early can be processed more efficiently in time and space.

When **SELECT IF** is specified following **TEMPORARY** (see Section 13.6 [TEMPORARY], page 122), the **LAG** function may not be used (see [LAG], page 57).

13.5 SPLIT FILE

SPLIT FILE [{LAYERED, SEPARATE}] BY *var_list*.
SPLIT FILE OFF.

SPLIT FILE allows multiple sets of data present in one data file to be analyzed separately using single statistical procedure commands.

Specify a list of variable names to analyze multiple sets of data separately. Groups of adjacent cases having the same values for these variables are analyzed by statistical procedure commands as one group. An independent analysis is carried out for each group of cases, and the variable values for the group are printed along with the analysis.

When a list of variable names is specified, one of the keywords LAYERED or SEPARATE may also be specified. If provided, either keyword are ignored.

Groups are formed only by *adjacent* cases. To create a split using a variable where like values are not adjacent in the working file, you should first sort the data by that variable (see Section 12.8 [SORT CASES], page 119).

Specify OFF to disable SPLIT FILE and resume analysis of the entire active dataset as a single group of data.

When SPLIT FILE is specified after TEMPORARY, it affects only the next procedure (see Section 13.6 [TEMPORARY], page 122).

13.6 TEMPORARY

TEMPORARY.

TEMPORARY is used to make the effects of transformations following its execution temporary. These transformations will affect only the execution of the next procedure or procedure-like command. Their effects will not be saved to the active dataset.

The only specification on TEMPORARY is the command name.

TEMPORARY may not appear within a DO IF or LOOP construct. It may appear only once between procedures and procedure-like commands.

Scratch variables cannot be used following TEMPORARY.

An example may help to clarify:

```
DATA LIST /X 1-2.
BEGIN DATA.
 2
 4
10
15
20
24
END DATA.

COMPUTE X=X/2.

TEMPORARY.
COMPUTE X=X+3.

DESCRIPTIVES X.
DESCRIPTIVES X.
```

The data read by the first DESCRIPTIVES are 4, 5, 8, 10.5, 13, 15. The data read by the first DESCRIPTIVES are 1, 2, 5, 7.5, 10, 12.

13.7 WEIGHT

WEIGHT BY *var_name*.
WEIGHT OFF.

WEIGHT assigns cases varying weights, changing the frequency distribution of the active dataset. Execution of WEIGHT is delayed until data have been read.

If a variable name is specified, WEIGHT causes the values of that variable to be used as weighting factors for subsequent statistical procedures. Use of keyword BY is optional but recommended. Weighting variables must be numeric. Scratch variables may not be used for weighting (see Section 6.7.5 [Scratch Variables], page 43).

When OFF is specified, subsequent statistical procedures will weight all cases equally.

A positive integer weighting factor w on a case will yield the same statistical output as would replicating the case w times. A weighting factor of 0 is treated for statistical purposes as if the case did not exist in the input. Weighting values need not be integers, but negative and system-missing values for the weighting variable are interpreted as weighting factors of 0. User-missing values are not treated specially.

When WEIGHT is specified after TEMPORARY, it affects only the next procedure (see Section 13.6 [TEMPORARY], page 122).

WEIGHT does not cause cases in the active dataset to be replicated in memory.

14 Conditional and Looping Constructs

This chapter documents PSPP commands used for conditional execution, looping, and flow of control.

14.1 BREAK

BREAK.

BREAK terminates execution of the innermost currently executing LOOP construct.

BREAK is allowed only inside LOOP...END LOOP. See Section 14.4 [LOOP], page 125, for more details.

14.2 DO IF

DO IF condition.
   . . .
[ELSE IF condition.
   . . .
]. . .
[ELSE.
   . . .]
END IF.

DO IF allows one of several sets of transformations to be executed, depending on user-specified conditions.

If the specified boolean expression evaluates as true, then the block of code following DO IF is executed. If it evaluates as missing, then none of the code blocks is executed. If it is false, then the boolean expression on the first ELSE IF, if present, is tested in turn, with the same rules applied. If all expressions evaluate to false, then the ELSE code block is executed, if it is present.

When DO IF or ELSE IF is specified following TEMPORARY (see Section 13.6 [TEMPORARY], page 122), the LAG function may not be used (see [LAG], page 57).

14.3 DO REPEAT

DO REPEAT dummy_name=expansion. . . .
   . . .
END REPEAT [PRINT].

expansion takes one of the following forms:
     var_list
     num_or_range. . .
     'string'. . .
     ALL

num_or_range takes one of the following forms:
     number
     num1 TO num2

DO REPEAT repeats a block of code, textually substituting different variables, numbers, or strings into the block with each repetition.

Specify a dummy variable name followed by an equals sign ('=') and the list of replacements. Replacements can be a list of existing or new variables, numbers, strings, or ALL to specify all existing variables. When numbers are specified, runs of increasing integers may be indicated as *num1* TO *num2*, so that '1 TO 5' is short for '1 2 3 4 5'.

Multiple dummy variables can be specified. Each variable must have the same number of replacements.

The code within DO REPEAT is repeated as many times as there are replacements for each variable. The first time, the first value for each dummy variable is substituted; the second time, the second value for each dummy variable is substituted; and so on.

Dummy variable substitutions work like macros. They take place anywhere in a line that the dummy variable name occurs. This includes command and subcommand names, so command and subcommand names that appear in the code block should not be used as dummy variable identifiers. Dummy variable substitutions do not occur inside quoted strings, comments, unquoted strings (such as the text on the TITLE or DOCUMENT command), or inside BEGIN DATA...END DATA.

Substitution occurs only on whole words, so that, for example, a dummy variable PRINT would not be substituted into the word PRINTOUT.

New variable names used as replacements are not automatically created as variables, but only if used in the code block in a context that would create them, e.g. on a NUMERIC or STRING command or on the left side of a COMPUTE assignment.

Any command may appear within DO REPEAT, including nested DO REPEAT commands. If INCLUDE or INSERT appears within DO REPEAT, the substitutions do not apply to the included file.

If PRINT is specified on END REPEAT, the commands after substitutions are made are printed to the listing file, prefixed by a plus sign ('+').

14.4 LOOP

> LOOP [*index_var*=*start* TO *end* [BY *incr*]] [IF *condition*].
>   . . .
> END LOOP [IF *condition*].

LOOP iterates a group of commands. A number of termination options are offered.

Specify index_var to make that variable count from one value to another by a particular increment. *index_var* must be a pre-existing numeric variable. *start*, *end*, and *incr* are numeric expressions (see Chapter 7 [Expressions], page 46.)

During the first iteration, *index_var* is set to the value of *start*. During each successive iteration, *index_var* is increased by the value of *incr*. If *end* > *start*, then the loop terminates when *index_var* > *end*; otherwise it terminates when *index_var* < *end*. If *incr* is not specified then it defaults to +1 or -1 as appropriate.

If *end* > *start* and *incr* < 0, or if *end* < *start* and *incr* > 0, then the loop is never executed. *index_var* is nevertheless set to the value of start.

Modifying *index_var* within the loop is allowed, but it has no effect on the value of *index_var* in the next iteration.

Specify a boolean expression for the condition on **LOOP** to cause the loop to be executed only if the condition is true. If the condition is false or missing before the loop contents are executed the first time, the loop contents are not executed at all.

If index and condition clauses are both present on **LOOP**, the index variable is always set before the condition is evaluated. Thus, a condition that makes use of the index variable will always see the index value to be used in the next execution of the body.

Specify a boolean expression for the condition on **END LOOP** to cause the loop to terminate if the condition is true after the enclosed code block is executed. The condition is evaluated at the end of the loop, not at the beginning, so that the body of a loop with only a condition on **END LOOP** will always execute at least once.

If neither the index clause nor either condition clause is present, then the loop is executed *max_loops* (see Section 16.20 [SET], page 157) times. The default value of *max_loops* is 40.

BREAK also terminates **LOOP** execution (see Section 14.1 [BREAK], page 124).

Loop index variables are by default reset to system-missing from one case to another, not left, unless a scratch variable is used as index. When loops are nested, this is usually undesired behavior, which can be corrected with **LEAVE** (see Section 11.5 [LEAVE], page 101) or by using a scratch variable as the loop index.

When **LOOP** or **END LOOP** is specified following **TEMPORARY** (see Section 13.6 [TEMPORARY], page 122), the **LAG** function may not be used (see [LAG], page 57).

15 Statistics

This chapter documents the statistical procedures that PSPP supports so far.

15.1 DESCRIPTIVES

```
DESCRIPTIVES
      /VARIABLES=var_list
      /MISSING={VARIABLE,LISTWISE} {INCLUDE,NOINCLUDE}
      /FORMAT={LABELS,NOLABELS} {NOINDEX,INDEX} {LINE,SERIAL}
      /SAVE
      /STATISTICS={ALL,MEAN,SEMEAN,STDDEV,VARIANCE,KURTOSIS,
            SKEWNESS,RANGE,MINIMUM,MAXIMUM,SUM,DEFAULT,
            SESKEWNESS,SEKURTOSIS}
      /SORT={NONE,MEAN,SEMEAN,STDDEV,VARIANCE,KURTOSIS,SKEWNESS,
         RANGE,MINIMUM,MAXIMUM,SUM,SESKEWNESS,SEKURTOSIS,NAME}
            {A,D}
```

The DESCRIPTIVES procedure reads the active dataset and outputs descriptive statistics requested by the user. In addition, it can optionally compute Z-scores.

The VARIABLES subcommand, which is required, specifies the list of variables to be analyzed. Keyword VARIABLES is optional.

All other subcommands are optional:

The MISSING subcommand determines the handling of missing variables. If INCLUDE is set, then user-missing values are included in the calculations. If NOINCLUDE is set, which is the default, user-missing values are excluded. If VARIABLE is set, then missing values are excluded on a variable by variable basis; if LISTWISE is set, then the entire case is excluded whenever any value in that case has a system-missing or, if INCLUDE is set, user-missing value.

The FORMAT subcommand affects the output format. Currently the LABELS/NOLABELS and NOINDEX/INDEX settings are not used. When SERIAL is set, both valid and missing number of cases are listed in the output; when NOSERIAL is set, only valid cases are listed.

The SAVE subcommand causes DESCRIPTIVES to calculate Z scores for all the specified variables. The Z scores are saved to new variables. Variable names are generated by trying first the original variable name with Z prepended and truncated to a maximum of 8 characters, then the names ZSC000 through ZSC999, STDZ00 through STDZ09, ZZZZ00 through ZZZZ09, ZQZQ00 through ZQZQ09, in that sequence. In addition, Z score variable names can be specified explicitly on VARIABLES in the variable list by enclosing them in parentheses after each variable. When Z scores are calculated, PSPP ignores TEMPORARY, treating temporary transformations as permanent.

The STATISTICS subcommand specifies the statistics to be displayed:

ALL   All of the statistics below.

MEAN   Arithmetic mean.

SEMEAN   Standard error of the mean.

STDDEV   Standard deviation.

VARIANCE   Variance.

KURTOSIS   Kurtosis and standard error of the kurtosis.

SKEWNESS   Skewness and standard error of the skewness.

RANGE   Range.

MINIMUM   Minimum value.

MAXIMUM   Maximum value.

SUM   Sum.

DEFAULT   Mean, standard deviation of the mean, minimum, maximum.

SEKURTOSIS
           Standard error of the kurtosis.

SESKEWNESS
           Standard error of the skewness.

The SORT subcommand specifies how the statistics should be sorted. Most of the possible values should be self-explanatory. NAME causes the statistics to be sorted by name. By default, the statistics are listed in the order that they are specified on the VARIABLES subcommand. The A and D settings request an ascending or descending sort order, respectively.

15.2 FREQUENCIES

```
FREQUENCIES
     /VARIABLES=var_list
     /FORMAT={TABLE,NOTABLE,LIMIT(limit)}
          {AVALUE,DVALUE,AFREQ,DFREQ}
     /MISSING={EXCLUDE,INCLUDE}
     /STATISTICS={DEFAULT,MEAN,SEMEAN,MEDIAN,MODE,STDDEV,VARIANCE,
          KURTOSIS,SKEWNESS,RANGE,MINIMUM,MAXIMUM,SUM,
          SESKEWNESS,SEKURTOSIS,ALL,NONE}
     /NTILES=ntiles
     /PERCENTILES=percent...
     /HISTOGRAM=[MINIMUM(x_min)] [MAXIMUM(x_max)]
          [{FREQ[(y_max)],PERCENT[(y_max)]}] [{NONORMAL,NORMAL}]
     /PIECHART=[MINIMUM(x_min)] [MAXIMUM(x_max)]
          [{FREQ,PERCENT}] [{NOMISSING,MISSING}]
     /BARCHART=[MINIMUM(x_min)] [MAXIMUM(x_max)]
          [{FREQ,PERCENT}]
     /ORDER={ANALYSIS,VARIABLE}


(These options are not currently implemented.)
     /HBAR=...
     /GROUPED=...
```

The FREQUENCIES procedure outputs frequency tables for specified variables. FREQUENCIES can also calculate and display descriptive statistics (including median and mode) and percentiles, and various graphical representations of the frequency distribution.

The `VARIABLES` subcommand is the only required subcommand. Specify the variables to be analyzed.

The `FORMAT` subcommand controls the output format. It has several possible settings:

`TABLE`, the default, causes a frequency table to be output for every variable specified. `NOTABLE` prevents them from being output. `LIMIT` with a numeric argument causes them to be output except when there are more than the specified number of values in the table.

Normally frequency tables are sorted in ascending order by value. This is `AVALUE`. `DVALUE` tables are sorted in descending order by value. `AFREQ` and `DFREQ` tables are sorted in ascending and descending order, respectively, by frequency count.

The `MISSING` subcommand controls the handling of user-missing values. When `EXCLUDE`, the default, is set, user-missing values are not included in frequency tables or statistics. When `INCLUDE` is set, user-missing are included. System-missing values are never included in statistics, but are listed in frequency tables.

The available `STATISTICS` are the same as available in `DESCRIPTIVES` (see Section 15.1 [DESCRIPTIVES], page 127), with the addition of `MEDIAN`, the data's median value, and `MODE`, the mode. (If there are multiple modes, the smallest value is reported.) By default, the mean, standard deviation of the mean, minimum, and maximum are reported for each variable.

`PERCENTILES` causes the specified percentiles to be reported. The percentiles should be presented at a list of numbers between 0 and 100 inclusive. The `NTILES` subcommand causes the percentiles to be reported at the boundaries of the data set divided into the specified number of ranges. For instance, `/NTILES=4` would cause quartiles to be reported.

The `HISTOGRAM` subcommand causes the output to include a histogram for each specified numeric variable. The X axis by default ranges from the minimum to the maximum value observed in the data, but the `MINIMUM` and `MAXIMUM` keywords can set an explicit range.[1] Histograms are not created for string variables.

Specify `NORMAL` to superimpose a normal curve on the histogram.

The `PIECHART` subcommand adds a pie chart for each variable to the data. Each slice represents one value, with the size of the slice proportional to the value's frequency. By default, all non-missing values are given slices. The `MINIMUM` and `MAXIMUM` keywords can be used to limit the displayed slices to a given range of values. The keyword `NOMISSING` causes missing values to be omitted from the piechart. This is the default. If instead, `MISSING` is specified, then a single slice will be included representing all system missing and user-missing cases.

The `BARCHART` subcommand produces a bar chart for each variable. The `MINIMUM` and `MAXIMUM` keywords can be used to omit categories whose counts which lie outside the specified limits. The `FREQ` option (default) causes the ordinate to display the frequency of each category, whereas the `PERCENT` option will display relative percentages.

The `FREQ` and `PERCENT` options on `HISTOGRAM` and `PIECHART` are accepted but not currently honoured.

---

[1] The number of bins is chosen according to the Freedman-Diaconis rule: $2 \times IQR(x)n^{-1/3}$, where $IQR(x)$ is the interquartile range of x and n is the number of samples. Note that `EXAMINE` uses a different algorithm to determine bin sizes.

The ORDER subcommand is accepted but ignored.

15.3 EXAMINE

```
EXAMINE
      VARIABLES= var1 [var2] ... [varN]
        [BY factor1 [BY subfactor1]
          [ factor2 [BY subfactor2]]
          ...
          [ factor3 [BY subfactor3]]
        ]
      /STATISTICS={DESCRIPTIVES, EXTREME[(n)], ALL, NONE}
    /PLOT={BOXPLOT, NPPLOT, HISTOGRAM, SPREADLEVEL[(t)], ALL, NONE}
      /CINTERVAL p
      /COMPARE={GROUPS,VARIABLES}
      /ID=identity_variable
      /{TOTAL,NOTOTAL}
      /PERCENTILE=[percentiles]={HAVERAGE, WAVERAGE, ROUND, AEM-
PIRICAL, EMPIRICAL }
      /MISSING={LISTWISE, PAIRWISE} [{EXCLUDE, INCLUDE}]
[{NOREPORT,REPORT}]
```

The EXAMINE command is used to perform exploratory data analysis. In particular, it is useful for testing how closely a distribution follows a normal distribution, and for finding outliers and extreme values.

The VARIABLES subcommand is mandatory. It specifies the dependent variables and optionally variables to use as factors for the analysis. Variables listed before the first BY keyword (if any) are the dependent variables. The dependent variables may optionally be followed by a list of factors which tell PSPP how to break down the analysis for each dependent variable.

Following the dependent variables, factors may be specified. The factors (if desired) should be preceded by a single BY keyword. The format for each factor is

   factorvar [BY subfactorvar].

Each unique combination of the values of factorvar and subfactorvar divide the dataset into cells. Statistics will be calculated for each cell and for the entire dataset (unless NOTOTAL is given).

The STATISTICS subcommand specifies which statistics to show. DESCRIPTIVES will produce a table showing some parametric and non-parametrics statistics. EXTREME produces a table showing the extremities of each cell. A number in parentheses, n determines how many upper and lower extremities to show. The default number is 5.

The subcommands TOTAL and NOTOTAL are mutually exclusive. If TOTAL appears, then statistics will be produced for the entire dataset as well as for each cell. If NOTOTAL appears, then statistics will be produced only for the cells (unless no factor variables have been given). These subcommands have no effect if there have been no factor variables specified.

The PLOT subcommand specifies which plots are to be produced if any. Available plots are HISTOGRAM, NPPLOT, BOXPLOT and SPREADLEVEL. The first three can be used to visualise how closely each cell conforms to a normal distribution, whilst the spread vs. level plot can be useful to visualise how the variance of differs between factors. Boxplots will also show you the outliers and extreme values.[2]

The SPREADLEVEL plot displays the interquartile range versus the median. It takes an optional parameter t, which specifies how the data should be transformed prior to plotting. The given value t is a power to which the data is raised. For example, if t is given as 2, then the data will be squared. Zero, however is a special value. If t is 0 or is omitted, then data will be transformed by taking its natural logarithm instead of raising to the power of t.

The COMPARE subcommand is only relevant if producing boxplots, and it is only useful there is more than one dependent variable and at least one factor. If /COMPARE=GROUPS is specified, then one plot per dependent variable is produced, each of which contain boxplots for all the cells. If /COMPARE=VARIABLES is specified, then one plot per cell is produced, each containing one boxplot per dependent variable. If the /COMPARE subcommand is omitted, then PSPP behaves as if /COMPARE=GROUPS were given.

The ID subcommand is relevant only if /PLOT=BOXPLOT or /STATISTICS=EXTREME has been given. If given, it should provide the name of a variable which is to be used to labels extreme values and outliers. Numeric or string variables are permissible. If the ID subcommand is not given, then the case number will be used for labelling.

The CINTERVAL subcommand specifies the confidence interval to use in calculation of the descriptives command. The default is 95%.

The PERCENTILES subcommand specifies which percentiles are to be calculated, and which algorithm to use for calculating them. The default is to calculate the 5, 10, 25, 50, 75, 90, 95 percentiles using the HAVERAGE algorithm.

The TOTAL and NOTOTAL subcommands are mutually exclusive. If NOTOTAL is given and factors have been specified in the VARIABLES subcommand, then then statistics for the unfactored dependent variables are produced in addition to the factored variables. If there are no factors specified then TOTAL and NOTOTAL have no effect.

The following example will generate descriptive statistics and histograms for two variables *score1* and *score2*. Two factors are given, *viz*: *gender* and *gender* BY *culture*. Therefore, the descriptives and histograms will be generated for each distinct value of *gender and* for each distinct combination of the values of *gender* and *race*. Since the NOTOTAL keyword is given, statistics and histograms for *score1* and *score2* covering the whole dataset are not produced.

```
EXAMINE score1 score2 BY
        gender
        gender BY culture
        /STATISTICS = DESCRIPTIVES
        /PLOT = HISTOGRAM
        /NOTOTAL.
```

Here is a second example showing how the **examine** command can be used to find extremities.

---

[2] HISTOGRAM uses Sturges' rule to determine the number of bins, as approximately $1 + \log 2(n)$, where n is the number of samples. Note that FREQUENCIES uses a different algorithm to find the bin size.

```
EXAMINE height weight BY
        gender
        /STATISTICS = EXTREME (3)
        /PLOT = BOXPLOT
        /COMPARE = GROUPS
        /ID = name.
```

In this example, we look at the height and weight of a sample of individuals and how they differ between male and female. A table showing the 3 largest and the 3 smallest values of *height* and *weight* for each gender, and for the whole dataset will be shown. Boxplots will also be produced. Because /COMPARE = GROUPS was given, boxplots for male and female will be shown in the same graphic, allowing us to easily see the difference between the genders. Since the variable *name* was specified on the ID subcommand, this will be used to label the extreme values.

Warning! If many dependent variables are specified, or if factor variables are specified for which there are many distinct values, then EXAMINE will produce a very large quantity of output.

15.4 GRAPH

```
GRAPH
    /HISTOGRAM = var
    /SCATTERPLOT [(BIVARIATE)] = var1 WITH var2 [BY var3]
    [ /MISSING={LISTWISE, VARIABLE} [{EXCLUDE, INCLUDE}] ]
[{NOREPORT,REPORT}]
```

The GRAPH produces graphical plots of data. Only one of the subcommands HISTOGRAM or SCATTERPLOT can be specified, i.e. only one plot can be produced per call of GRAPH. The MISSING is optional.

The subcommand SCATTERPLOT produces an xy plot of the data. The different values of the optional third variable *var3* will result in different colours and/or markers for the plot. The following is an example for producing a scatterplot.

```
GRAPH
    /SCATTERPLOT = height WITH weight BY gender.
```

This example will produce a scatterplot where *height* is plotted versus *weight*. Depending on the value of the *gender* variable, the colour of the datapoint is different. With this plot it is possible to analyze gender differences for *height* vs. *weight* relation.

The subcommand HISTOGRAM produces a histogram. Only one variable is allowed for the histogram plot. For an alternative method to produce histograms see Section 15.3 [EXAMINE], page 130. The following example produces a histogram plot for the variable *weight*.

```
GRAPH
    /HISTOGRAM = weight.
```

15.5 CORRELATIONS

```
CORRELATIONS
```

/VARIABLES = *var_list* [WITH *var_list*]
[
  .
  .
  .
/VARIABLES = *var_list* [WITH *var_list*]
/VARIABLES = *var_list* [WITH *var_list*]
]

[/PRINT={TWOTAIL, ONETAIL} {SIG, NOSIG}]
[/STATISTICS=DESCRIPTIVES XPROD ALL]
[/MISSING={PAIRWISE, LISTWISE} {INCLUDE, EXCLUDE}]

The CORRELATIONS procedure produces tables of the Pearson correlation coefficient for a set of variables. The significance of the coefficients are also given.

At least one VARIABLES subcommand is required. If the WITH keyword is used, then a non-square correlation table will be produced. The variables preceding WITH, will be used as the rows of the table, and the variables following will be the columns of the table. If no WITH subcommand is given, then a square, symmetrical table using all variables is produced.

The MISSING subcommand determines the handling of missing variables. If INCLUDE is set, then user-missing values are included in the calculations, but system-missing values are not. If EXCLUDE is set, which is the default, user-missing values are excluded as well as system-missing values.

If LISTWISE is set, then the entire case is excluded from analysis whenever any variable specified in any /VARIABLES subcommand contains a missing value. If PAIRWISE is set, then a case is considered missing only if either of the values for the particular coefficient are missing. The default is PAIRWISE.

The PRINT subcommand is used to control how the reported significance values are printed. If the TWOTAIL option is used, then a two-tailed test of significance is printed. If the ONETAIL option is given, then a one-tailed test is used. The default is TWOTAIL.

If the NOSIG option is specified, then correlation coefficients with significance less than 0.05 are highlighted. If SIG is specified, then no highlighting is performed. This is the default.

The STATISTICS subcommand requests additional statistics to be displayed. The keyword DESCRIPTIVES requests that the mean, number of non-missing cases, and the non-biased estimator of the standard deviation are displayed. These statistics will be displayed in a separated table, for all the variables listed in any /VARIABLES subcommand. The XPROD keyword requests cross-product deviations and covariance estimators to be displayed for each pair of variables. The keyword ALL is the union of DESCRIPTIVES and XPROD.

15.6 CROSSTABS

CROSSTABS
    /TABLES=*var_list* BY *var_list* [BY *var_list*]...
    /MISSING={TABLE,INCLUDE,REPORT}
    /WRITE={NONE,CELLS,ALL}
    /FORMAT={TABLES,NOTABLES}

```
                   {PIVOT,NOPIVOT}
                   {AVALUE,DVALUE}
                   {NOINDEX,INDEX}
                   {BOX,NOBOX}
            /CELLS={COUNT,ROW,COLUMN,TOTAL,EXPECTED,RESIDUAL,SRESIDUAL,
                   ASRESIDUAL,ALL,NONE}
            /STATISTICS={CHISQ,PHI,CC,LAMBDA,UC,BTAU,CTAU,RISK,GAMMA,D,
                   KAPPA,ETA,CORR,ALL,NONE}
              /BARCHART
```

(Integer mode.)

    /VARIABLES=*var_list* (*low,high*)...

The `CROSSTABS` procedure displays crosstabulation tables requested by the user. It can calculate several statistics for each cell in the crosstabulation tables. In addition, a number of statistics can be calculated for each table itself.

The `TABLES` subcommand is used to specify the tables to be reported. Any number of dimensions is permitted, and any number of variables per dimension is allowed. The `TABLES` subcommand may be repeated as many times as needed. This is the only required subcommand in *general mode*.

Occasionally, one may want to invoke a special mode called *integer mode*. Normally, in general mode, PSPP automatically determines what values occur in the data. In integer mode, the user specifies the range of values that the data assumes. To invoke this mode, specify the `VARIABLES` subcommand, giving a range of data values in parentheses for each variable to be used on the `TABLES` subcommand. Data values inside the range are truncated to the nearest integer, then assigned to that value. If values occur outside this range, they are discarded. When it is present, the `VARIABLES` subcommand must precede the `TABLES` subcommand.

In general mode, numeric and string variables may be specified on TABLES. In integer mode, only numeric variables are allowed.

The `MISSING` subcommand determines the handling of user-missing values. When set to `TABLE`, the default, missing values are dropped on a table by table basis. When set to `INCLUDE`, user-missing values are included in tables and statistics. When set to `REPORT`, which is allowed only in integer mode, user-missing values are included in tables but marked with an 'M' (for "missing") and excluded from statistical calculations.

Currently the `WRITE` subcommand is ignored.

The `FORMAT` subcommand controls the characteristics of the crosstabulation tables to be displayed. It has a number of possible settings:

    `TABLES`, the default, causes crosstabulation tables to be output. `NOTABLES` suppresses them.

    `PIVOT`, the default, causes each `TABLES` subcommand to be displayed in a pivot table format. `NOPIVOT` causes the old-style crosstabulation format to be used.

    `AVALUE`, the default, causes values to be sorted in ascending order. `DVALUE` asserts a descending sort order.

    `INDEX` and `NOINDEX` are currently ignored.

BOX and NOBOX is currently ignored.

The CELLS subcommand controls the contents of each cell in the displayed crosstabulation table. The possible settings are:

COUNT   Frequency count.

ROW   Row percent.

COLUMN   Column percent.

TOTAL   Table percent.

EXPECTED
        Expected value.

RESIDUAL
        Residual.

SRESIDUAL
        Standardized residual.

ASRESIDUAL
        Adjusted standardized residual.

ALL   All of the above.

NONE   Suppress cells entirely.

'/CELLS' without any settings specified requests COUNT, ROW, COLUMN, and TOTAL. If CELLS is not specified at all then only COUNT will be selected.

The STATISTICS subcommand selects statistics for computation:

CHISQ
        Pearson chi-square, likelihood ratio, Fisher's exact test, continuity correction, linear-by-linear association.

PHI   Phi.

CC   Contingency coefficient.

LAMBDA   Lambda.

UC   Uncertainty coefficient.

BTAU   Tau-b.

CTAU   Tau-c.

RISK   Risk estimate.

GAMMA   Gamma.

D   Somers' D.

KAPPA   Cohen's Kappa.

ETA   Eta.

CORR   Spearman correlation, Pearson's r.

ALL   All of the above.

NONE   No statistics.

   Selected statistics are only calculated when appropriate for the statistic. Certain statistics require tables of a particular size, and some statistics are calculated only in integer mode.

   '/STATISTICS' without any settings selects CHISQ. If the STATISTICS subcommand is not given, no statistics are calculated.

   The '/BARCHART' subcommand produces a clustered bar chart for the first two variables on each table. If a table has more than two variables, the counts for the third and subsequent levels will be aggregated and the chart will be produces as if there were only two variables.

   Please note: Currently the implementation of CROSSTABS has the following limitations:

- Significance of some symmetric and directional measures is not calculated.

- Asymptotic standard error is not calculated for Goodman and Kruskal's tau or symmetric Somers' d.

- Approximate T is not calculated for symmetric uncertainty coefficient.

   Fixes for any of these deficiencies would be welcomed.

15.7 FACTOR

   FACTOR VARIABLES=var_list

      [/METHOD = {CORRELATION, COVARIANCE}]

      [/ANALYSIS=var_list]

      [/EXTRACTION={PC, PAF}]

     [/ROTATION={VARIMAX, EQUAMAX, QUARTIMAX, PROMAX[(k)], NOROTATE}]

      [/PRINT=[INITIAL] [EXTRACTION] [ROTATION] [UNIVARIATE] [COR-RELATION] [COVARIANCE] [DET] [KMO] [SIG] [ALL] [DEFAULT]]

      [/PLOT=[EIGEN]]

      [/FORMAT=[SORT] [BLANK(n)] [DEFAULT]]

       [/CRITERIA=[FACTORS(n)] [MINEIGEN(l)] [ITERATE(m)] [ECONVERGE ($delta$)] [DEFAULT]]

      [/MISSING=[{LISTWISE, PAIRWISE}] [{INCLUDE, EXCLUDE}]]

   The FACTOR command performs Factor Analysis or Principal Axis Factoring on a dataset. It may be used to find common factors in the data or for data reduction purposes.

   The VARIABLES subcommand is required. It lists the variables which are to partake in the analysis. (The ANALYSIS subcommand may optionally further limit the variables that participate; it is not useful and implemented only for compatibility.)

The /EXTRACTION subcommand is used to specify the way in which factors (components) are extracted from the data. If PC is specified, then Principal Components Analysis is used. If PAF is specified, then Principal Axis Factoring is used. By default Principal Components Analysis will be used.

The /ROTATION subcommand is used to specify the method by which the extracted solution will be rotated. Three orthogonal rotation methods are available: VARIMAX (which is the default), EQUAMAX, and QUARTIMAX. There is one oblique rotation method, *viz*: PROMAX. Optionally you may enter the power of the promax rotation k, which must be enclosed in parentheses. The default value of k is 5. If you don't want any rotation to be performed, the word NOROTATE will prevent the command from performing any rotation on the data.

The /METHOD subcommand should be used to determine whether the covariance matrix or the correlation matrix of the data is to be analysed. By default, the correlation matrix is analysed.

The /PRINT subcommand may be used to select which features of the analysis are reported:

- UNIVARIATE A table of mean values, standard deviations and total weights are printed.
- INITIAL Initial communalities and eigenvalues are printed.
- EXTRACTION Extracted communalities and eigenvalues are printed.
- ROTATION Rotated communalities and eigenvalues are printed.
- CORRELATION The correlation matrix is printed.
- COVARIANCE The covariance matrix is printed.
- DET The determinant of the correlation or covariance matrix is printed.
- KMO The Kaiser-Meyer-Olkin measure of sampling adequacy and the Bartlett test of sphericity is printed.
- SIG The significance of the elements of correlation matrix is printed.
- ALL All of the above are printed.
- DEFAULT Identical to INITIAL and EXTRACTION.

If /PLOT=EIGEN is given, then a "Scree" plot of the eigenvalues will be printed. This can be useful for visualizing which factors (components) should be retained.

The /FORMAT subcommand determined how data are to be displayed in loading matrices. If SORT is specified, then the variables are sorted in descending order of significance. If BLANK(n) is specified, then coefficients whose absolute value is less than n will not be printed. If the keyword DEFAULT is given, or if no /FORMAT subcommand is given, then no sorting is performed, and all coefficients will be printed.

The /CRITERIA subcommand is used to specify how the number of extracted factors (components) are chosen. If FACTORS(n) is specified, where n is an integer, then n factors will be extracted. Otherwise, the MINEIGEN setting will be used. MINEIGEN(l) requests that all factors whose eigenvalues are greater than or equal to l are extracted. The default value of l is 1. The ECONVERGE setting has effect only when iterative algorithms for factor extraction (such as Principal Axis Factoring) are used. ECONVERGE(*delta*) specifies that iteration should cease when the maximum absolute value of the communality estimate between one iteration and the previous is less than *delta*. The default value of *delta* is 0.001. The ITERATE(m) may appear any number of times and is used for two different purposes.

It is used to set the maximum number of iterations (m) for convergence and also to set the maximum number of iterations for rotation. Whether it affects convergence or rotation depends upon which subcommand follows the ITERATE subcommand. If EXTRACTION follows, it affects convergence. If ROTATION follows, it affects rotation. If neither ROTATION nor EXTRACTION follow a ITERATE subcommand it will be ignored. The default value of m is 25.

The MISSING subcommand determines the handling of missing variables. If INCLUDE is set, then user-missing values are included in the calculations, but system-missing values are not. If EXCLUDE is set, which is the default, user-missing values are excluded as well as system-missing values. This is the default. If LISTWISE is set, then the entire case is excluded from analysis whenever any variable specified in the VARIABLES subcommand contains a missing value. If PAIRWISE is set, then a case is considered missing only if either of the values for the particular coefficient are missing. The default is LISTWISE.

15.8 LOGISTIC REGRESSION

LOGISTIC REGRESSION [VARIABLES =] *dependent_var* WITH *predictors*

[/CATEGORICAL = *categorical_predictors*]

[{/NOCONST | /ORIGIN | /NOORIGIN }]

[/PRINT = [SUMMARY] [DEFAULT] [CI(*confidence*)] [ALL]]

[/CRITERIA = [BCON(*min_delta*)] [ITERATE(*max_interations*)]
        [LCON(*min_likelihood_delta*)] [EPS(*min_epsilon*)]
        [CUT(*cut_point*)]]

[/MISSING = {INCLUDE|EXCLUDE}]

Bivariate Logistic Regression is used when you want to explain a dichotomous dependent variable in terms of one or more predictor variables.

The minimum command is

```
LOGISTIC REGRESSION y WITH x1 x2 ... xn.
```

Here, *y* is the dependent variable, which must be dichotomous and *x1* ... *xn* are the predictor variables whose coefficients the procedure estimates.

By default, a constant term is included in the model. Hence, the full model is $\mathbf{y} = b_0 + b_1\mathbf{x_1} + b_2\mathbf{x_2} + \ldots + b_n\mathbf{x_n}$

Predictor variables which are categorical in nature should be listed on the /CATEGORICAL subcommand. Simple variables as well as interactions between variables may be listed here.

If you want a model without the constant term b_0, use the keyword /ORIGIN. /NOCONST is a synonym for /ORIGIN.

An iterative Newton-Raphson procedure is used to fit the model. The /CRITERIA subcommand is used to specify the stopping criteria of the procedure, and other parameters. The value of *cut_point* is used in the classification table. It is the threshold above which predicted values are considered to be 1. Values of *cut_point* must lie in the range [0,1]. During iterations, if any one of the stopping criteria are satisfied, the procedure is considered complete. The stopping criteria are:

- The number of iterations exceeds *max_iterations*. The default value of *max_iterations* is 20.

- The change in the all coefficient estimates are less than *min_delta*. The default value of *min_delta* is 0.001.

- The magnitude of change in the likelihood estimate is less than *min_likelihood_delta*. The default value of *min_delta* is zero. This means that this criterion is disabled.

- The differential of the estimated probability for all cases is less than *min_epsilon*. In other words, the probabilities are close to zero or one. The default value of *min_epsilon* is 0.00000001.

The **PRINT** subcommand controls the display of optional statistics. Currently there is one such option, **CI**, which indicates that the confidence interval of the odds ratio should be displayed as well as its value. **CI** should be followed by an integer in parentheses, to indicate the confidence level of the desired confidence interval.

The **MISSING** subcommand determines the handling of missing variables. If **INCLUDE** is set, then user-missing values are included in the calculations, but system-missing values are not. If **EXCLUDE** is set, which is the default, user-missing values are excluded as well as system-missing values. This is the default.

15.9 MEANS

```
MEANS [TABLES =]
      {var_list}
        [ BY {var_list} [BY {var_list} [BY {var_list} ... ]]]

      [ /{var_list}
        [ BY {var_list} [BY {var_list} [BY {var_list} ... ]]] ]

    [/CELLS = [MEAN] [COUNT] [STDDEV] [SEMEAN] [SUM] [MIN] [MAX] [RANGE]
        [VARIANCE] [KURT] [SEKURT]
        [SKEW] [SESKEW] [FIRST] [LAST]
        [HARMONIC] [GEOMETRIC]
        [DEFAULT]
        [ALL]
        [NONE] ]

    [/MISSING = [TABLE] [INCLUDE] [DEPENDENT]]]
```

You can use the **MEANS** command to calculate the arithmetic mean and similar statistics, either for the dataset as a whole or for categories of data.

The simplest form of the command is

```
MEANS v.
```

which calculates the mean, count and standard deviation for *v*. If you specify a grouping variable, for example

```
    MEANS v BY g.
```
then the means, counts and standard deviations for v after having been grouped by g will be calculated. Instead of the mean, count and standard deviation, you could specify the statistics in which you are interested:
```
    MEANS x y BY g
          /CELLS = HARMONIC SUM MIN.
```
This example calculates the harmonic mean, the sum and the minimum values of x and y grouped by g.

The `CELLS` subcommand specifies which statistics to calculate. The available statistics are:

- `MEAN` The arithmetic mean.
- `COUNT` The count of the values.
- `STDDEV` The standard deviation.
- `SEMEAN` The standard error of the mean.
- `SUM` The sum of the values.
- `MIN` The minimum value.
- `MAX` The maximum value.
- `RANGE` The difference between the maximum and minimum values.
- `VARIANCE` The variance.
- `FIRST` The first value in the category.
- `LAST` The last value in the category.
- `SKEW` The skewness.
- `SESKEW` The standard error of the skewness.
- `KURT` The kurtosis
- `SEKURT` The standard error of the kurtosis.
- `HARMONIC` The harmonic mean.
- `GEOMETRIC` The geometric mean.

In addition, three special keywords are recognized:

- `DEFAULT` This is the same as `MEAN COUNT STDDEV`.
- `ALL` All of the above statistics will be calculated.
- `NONE` No statistics will be calculated (only a summary will be shown).

More than one *table* can be specified in a single command. Each table is separated by a '/'. For example
```
    MEANS TABLES =
          c d e BY x
          /a b BY x y
          /f BY y BY z.
```
has three tables (the '`TABLE =`' is optional). The first table has three dependent variables c, d and e and a single categorical variable x. The second table has two dependent variables a and b, and two categorical variables x and y. The third table has a single dependent variables f and a categorical variable formed by the combination of y and z.

By default values are omitted from the analysis only if missing values (either system missing or user missing) for any of the variables directly involved in their calculation are encountered. This behaviour can be modified with the /MISSING subcommand. Three options are possible: TABLE, INCLUDE and DEPENDENT.

/MISSING = TABLE causes cases to be dropped if any variable is missing in the table specification currently being processed, regardless of whether it is needed to calculate the statistic.

/MISSING = INCLUDE says that user missing values, either in the dependent variables or in the categorical variables should be taken at their face value, and not excluded.

/MISSING = DEPENDENT says that user missing values, in the dependent variables should be taken at their face value, however cases which have user missing values for the categorical variables should be omitted from the calculation.

15.10 NPAR TESTS

NPAR TESTS

nonparametric test subcommands
.
.
.

[/STATISTICS={DESCRIPTIVES}]

[/MISSING={ANALYSIS, LISTWISE} {INCLUDE, EXCLUDE}]

[/METHOD=EXACT [TIMER [(n)]]]

NPAR TESTS performs nonparametric tests. Non parametric tests make very few assumptions about the distribution of the data. One or more tests may be specified by using the corresponding subcommand. If the /STATISTICS subcommand is also specified, then summary statistics are produces for each variable that is the subject of any test.

Certain tests may take a long time to execute, if an exact figure is required. Therefore, by default asymptotic approximations are used unless the subcommand /METHOD=EXACT is specified. Exact tests give more accurate results, but may take an unacceptably long time to perform. If the TIMER keyword is used, it sets a maximum time, after which the test will be abandoned, and a warning message printed. The time, in minutes, should be specified in parentheses after the TIMER keyword. If the TIMER keyword is given without this figure, then a default value of 5 minutes is used.

15.10.1 Binomial test

[/BINOMIAL[(p)]=var_list[(value1[, value2)]]]

The /BINOMIAL subcommand compares the observed distribution of a dichotomous variable with that of a binomial distribution. The variable p specifies the test proportion of the binomial distribution. The default value of 0.5 is assumed if p is omitted.

If a single value appears after the variable list, then that value is used as the threshold to partition the observed values. Values less than or equal to the threshold value form the first category. Values greater than the threshold form the second category.

If two values appear after the variable list, then they will be used as the values which a variable must take to be in the respective category. Cases for which a variable takes a value equal to neither of the specified values, take no part in the test for that variable.

If no values appear, then the variable must assume dichotomous values. If more than two distinct, non-missing values for a variable under test are encountered then an error occurs.

If the test proportion is equal to 0.5, then a two tailed test is reported. For any other test proportion, a one tailed test is reported. For one tailed tests, if the test proportion is less than or equal to the observed proportion, then the significance of observing the observed proportion or more is reported. If the test proportion is more than the observed proportion, then the significance of observing the observed proportion or less is reported. That is to say, the test is always performed in the observed direction.

PSPP uses a very precise approximation to the gamma function to compute the binomial significance. Thus, exact results are reported even for very large sample sizes.

15.10.2 Chisquare Test

[/CHISQUARE=var_list[(lo,hi)] [/EXPECTED={EQUAL|f1, f2 ... fn}]]

The /CHISQUARE subcommand produces a chi-square statistic for the differences between the expected and observed frequencies of the categories of a variable. Optionally, a range of values may appear after the variable list. If a range is given, then non integer values are truncated, and values outside the specified range are excluded from the analysis.

The /EXPECTED subcommand specifies the expected values of each category. There must be exactly one non-zero expected value, for each observed category, or the EQUAL keyword must be specified. You may use the notation n*f to specify n consecutive expected categories all taking a frequency of f. The frequencies given are proportions, not absolute frequencies. The sum of the frequencies need not be 1. If no /EXPECTED subcommand is given, then then equal frequencies are expected.

15.10.3 Cochran Q Test

[/COCHRAN = var_list]

The Cochran Q test is used to test for differences between three or more groups. The data for var_list in all cases must assume exactly two distinct values (other than missing values).

The value of Q will be displayed and its Asymptotic significance based on a chi-square distribution.

15.10.4 Friedman Test

[/FRIEDMAN = var_list]

The Friedman test is used to test for differences between repeated measures when there is no indication that the distributions are normally distributed.

A list of variables which contain the measured data must be given. The procedure prints the sum of ranks for each variable, the test statistic and its significance.

15.10.5 Kendall's W Test

[/KENDALL = var_list]

The Kendall test investigates whether an arbitrary number of related samples come from the same population. It is identical to the Friedman test except that the additional statistic W, Kendall's Coefficient of Concordance is printed. It has the range [0,1] — a value of zero indicates no agreement between the samples whereas a value of unity indicates complete agreement.

15.10.6 Kolmogorov-Smirnov Test

[/KOLMOGOROV-SMIRNOV ({NORMAL [mu, sigma], UNIFORM [min, max], POIS-SON [lambda], EXPONENTIAL [scale] }) = var_list]

The one sample Kolmogorov-Smirnov subcommand is used to test whether or not a dataset is drawn from a particular distribution. Four distributions are supported, *viz:* Normal, Uniform, Poisson and Exponential.

Ideally you should provide the parameters of the distribution against which you wish to test the data. For example, with the normal distribution the mean (*mu*)and standard deviation (*sigma*) should be given; with the uniform distribution, the minimum (*min*)and maximum (*max*) value should be provided. However, if the parameters are omitted they will be imputed from the data. Imputing the parameters reduces the power of the test so should be avoided if possible.

In the following example, two variables *score* and *age* are tested to see if they follow a normal distribution with a mean of 3.5 and a standard deviation of 2.0.

```
NPAR TESTS
    /KOLMOGOROV-SMIRNOV (normal 3.5 2.0) = score age.
```

If the variables need to be tested against different distributions, then a separate subcommand must be used. For example the following syntax tests *score* against a normal distribution with mean of 3.5 and standard deviation of 2.0 whilst *age* is tested against a normal distribution of mean 40 and standard deviation 1.5.

```
NPAR TESTS
    /KOLMOGOROV-SMIRNOV (normal 3.5 2.0) = score
    /KOLMOGOROV-SMIRNOV (normal 40 1.5) =  age.
```

The abbreviated subcommand K-S may be used in place of KOLMOGOROV-SMIRNOV.

15.10.7 Kruskal-Wallis Test

[/KRUSKAL-WALLIS = var_list BY var (lower, upper)]

The Kruskal-Wallis test is used to compare data from an arbitrary number of populations. It does not assume normality. The data to be compared are specified by *var_list*. The categorical variable determining the groups to which the data belongs is given by *var*. The limits *lower* and *upper* specify the valid range of *var*. Any cases for which *var* falls outside [*lower*, *upper*] will be ignored.

The mean rank of each group as well as the chi-squared value and significance of the test will be printed. The abbreviated subcommand K-W may be used in place of KRUSKAL-WALLIS.

15.10.8 Mann-Whitney U Test

[/MANN-WHITNEY = var_list BY var (group1, group2)]

The Mann-Whitney subcommand is used to test whether two groups of data come from different populations. The variables to be tested should be specified in *var_list* and the grouping variable, that determines to which group the test variables belong, in *var*. *Var* may be either a string or an alpha variable. *Group1* and *group2* specify the two values of *var* which determine the groups of the test data. Cases for which the *var* value is neither *group1* or *group2* will be ignored.

The value of the Mann-Whitney U statistic, the Wilcoxon W, and the significance will be printed. The abbreviated subcommand M-W may be used in place of MANN-WHITNEY.

15.10.9 McNemar Test

[/MCNEMAR *var_list* [WITH *var_list* [(PAIRED)]]]

Use McNemar's test to analyse the significance of the difference between pairs of correlated proportions.

If the WITH keyword is omitted, then tests for all combinations of the listed variables are performed. If the WITH keyword is given, and the (PAIRED) keyword is also given, then the number of variables preceding WITH must be the same as the number following it. In this case, tests for each respective pair of variables are performed. If the WITH keyword is given, but the (PAIRED) keyword is omitted, then tests for each combination of variable preceding WITH against variable following WITH are performed.

The data in each variable must be dichotomous. If there are more than two distinct variables an error will occur and the test will not be run.

15.10.10 Median Test

[/MEDIAN [(value)] = var_list BY *variable* (value1, value2)]

The median test is used to test whether independent samples come from populations with a common median. The median of the populations against which the samples are to be tested may be given in parentheses immediately after the /MEDIAN subcommand. If it is not given, the median will be imputed from the union of all the samples.

The variables of the samples to be tested should immediately follow the '=' sign. The keyword BY must come next, and then the grouping variable. Two values in parentheses should follow. If the first value is greater than the second, then a 2 sample test is performed using these two values to determine the groups. If however, the first variable is less than the second, then a k sample test is conducted and the group values used are all values encountered which lie in the range [*value1,value2*].

15.10.11 Runs Test

[/RUNS ({MEAN, MEDIAN, MODE, value}) = var_list]

The /RUNS subcommand tests whether a data sequence is randomly ordered.

It works by examining the number of times a variable's value crosses a given threshold. The desired threshold must be specified within parentheses. It may either be specified as a number or as one of MEAN, MEDIAN or MODE. Following the threshold specification comes the list of variables whose values are to be tested.

The subcommand shows the number of runs, the asymptotic significance based on the length of the data.

15.10.12 Sign Test

[/SIGN *var_list* [WITH *var_list* [(PAIRED)]]]

The /SIGN subcommand tests for differences between medians of the variables listed. The test does not make any assumptions about the distribution of the data.

If the WITH keyword is omitted, then tests for all combinations of the listed variables are performed. If the WITH keyword is given, and the (PAIRED) keyword is also given, then the number of variables preceding WITH must be the same as the number following it. In this case, tests for each respective pair of variables are performed. If the WITH keyword is given, but the (PAIRED) keyword is omitted, then tests for each combination of variable preceding WITH against variable following WITH are performed.

15.10.13 Wilcoxon Matched Pairs Signed Ranks Test

[/WILCOXON *var_list* [WITH *var_list* [(PAIRED)]]]

The /WILCOXON subcommand tests for differences between medians of the variables listed. The test does not make any assumptions about the variances of the samples. It does however assume that the distribution is symmetrical.

If the WITH keyword is omitted, then tests for all combinations of the listed variables are performed. If the WITH keyword is given, and the (PAIRED) keyword is also given, then the number of variables preceding WITH must be the same as the number following it. In this case, tests for each respective pair of variables are performed. If the WITH keyword is given, but the (PAIRED) keyword is omitted, then tests for each combination of variable preceding WITH against variable following WITH are performed.

15.11 T-TEST

```
T-TEST
      /MISSING={ANALYSIS,LISTWISE} {EXCLUDE,INCLUDE}
      /CRITERIA=CIN(confidence)


(One Sample mode.)
      TESTVAL=test_value
      /VARIABLES=var_list


(Independent Samples mode.)
      GROUPS=var(value1 [, value2])
      /VARIABLES=var_list


(Paired Samples mode.)
      PAIRS=var_list [WITH var_list [(PAIRED)] ]
```

The `T-TEST` procedure outputs tables used in testing hypotheses about means. It operates in one of three modes:

- One Sample mode.
- Independent Groups mode.
- Paired mode.

Each of these modes are described in more detail below. There are two optional subcommands which are common to all modes.

The `/CRITERIA` subcommand tells PSPP the confidence interval used in the tests. The default value is 0.95.

The `MISSING` subcommand determines the handling of missing variables. If `INCLUDE` is set, then user-missing values are included in the calculations, but system-missing values are not. If `EXCLUDE` is set, which is the default, user-missing values are excluded as well as system-missing values. This is the default.

If `LISTWISE` is set, then the entire case is excluded from analysis whenever any variable specified in the `/VARIABLES`, `/PAIRS` or `/GROUPS` subcommands contains a missing value. If `ANALYSIS` is set, then missing values are excluded only in the analysis for which they would be needed. This is the default.

15.11.1 One Sample Mode

The `TESTVAL` subcommand invokes the One Sample mode. This mode is used to test a population mean against a hypothesized mean. The value given to the `TESTVAL` subcommand is the value against which you wish to test. In this mode, you must also use the `/VARIABLES` subcommand to tell PSPP which variables you wish to test.

15.11.2 Independent Samples Mode

The `GROUPS` subcommand invokes Independent Samples mode or 'Groups' mode. This mode is used to test whether two groups of values have the same population mean. In this mode, you must also use the `/VARIABLES` subcommand to tell PSPP the dependent variables you wish to test.

The variable given in the `GROUPS` subcommand is the independent variable which determines to which group the samples belong. The values in parentheses are the specific values of the independent variable for each group. If the parentheses are omitted and no values are given, the default values of 1.0 and 2.0 are assumed.

If the independent variable is numeric, it is acceptable to specify only one value inside the parentheses. If you do this, cases where the independent variable is greater than or equal to this value belong to the first group, and cases less than this value belong to the second group. When using this form of the `GROUPS` subcommand, missing values in the independent variable are excluded on a listwise basis, regardless of whether `/MISSING=LISTWISE` was specified.

15.11.3 Paired Samples Mode

The `PAIRS` subcommand introduces Paired Samples mode. Use this mode when repeated measures have been taken from the same samples. If the `WITH` keyword is omitted, then tables for all combinations of variables given in the `PAIRS` subcommand are generated. If the WITH keyword is given, and the (PAIRED) keyword is also given, then the number of variables preceding WITH must be the same as the number following it. In this case, tables for each respective pair of variables are generated. In the event that the WITH keyword is given, but the (PAIRED) keyword is omitted, then tables for each combination of variable preceding WITH against variable following WITH are generated.

15.12 ONEWAY

        ONEWAY
                [/VARIABLES =] var_list BY var
                /MISSING={ANALYSIS,LISTWISE} {EXCLUDE,INCLUDE}
                /CONTRAST= value1 [, value2] ... [,valueN]
                /STATISTICS={DESCRIPTIVES,HOMOGENEITY}
                /POSTHOC={BONFERRONI, GH, LSD, SCHEFFE, SIDAK, TUKEY, AL-
        PHA ([value])}

The ONEWAY procedure performs a one-way analysis of variance of variables factored by a single independent variable. It is used to compare the means of a population divided into more than two groups.

The dependent variables to be analysed should be given in the VARIABLES subcommand. The list of variables must be followed by the BY keyword and the name of the independent (or factor) variable.

You can use the STATISTICS subcommand to tell PSPP to display ancillary information. The options accepted are:

- DESCRIPTIVES Displays descriptive statistics about the groups factored by the independent variable.

- HOMOGENEITY Displays the Levene test of Homogeneity of Variance for the variables and their groups.

The CONTRAST subcommand is used when you anticipate certain differences between the groups. The subcommand must be followed by a list of numerals which are the coefficients of the groups to be tested. The number of coefficients must correspond to the number of distinct groups (or values of the independent variable). If the total sum of the coefficients are not zero, then PSPP will display a warning, but will proceed with the analysis. The CONTRAST subcommand may be given up to 10 times in order to specify different contrast tests. The MISSING subcommand defines how missing values are handled. If LISTWISE is specified then cases which have missing values for the independent variable or any dependent variable will be ignored. If ANALYSIS is specified, then cases will be ignored if the independent variable is missing or if the dependent variable currently being analysed is missing. The default is ANALYSIS. A setting of EXCLUDE means that variables whose values are user-missing are to be excluded from the analysis. A setting of INCLUDE means they are to be included. The default is EXCLUDE.

Using the POSTHOC subcommand you can perform multiple pairwise comparisons on the data. The following comparison methods are available:

- LSD Least Significant Difference.

- TUKEY Tukey Honestly Significant Difference.

- BONFERRONI Bonferroni test.

- SCHEFFE Scheffé's test.
- SIDAK Sidak test.
- GH The Games-Howell test.

The optional syntax ALPHA(*value*) is used to indicate that *value* should be used as the confidence level for which the posthoc tests will be performed. The default is 0.05.

15.13 QUICK CLUSTER

```
QUICK CLUSTER var_list
    [/CRITERIA=CLUSTERS(k) [MXITER(max_iter)]]
    [/MISSING={EXCLUDE,INCLUDE} {LISTWISE, PAIRWISE}]
```

The QUICK CLUSTER command performs k-means clustering on the dataset. This is useful when you wish to allocate cases into clusters of similar values and you already know the number of clusters.

The minimum specification is 'QUICK CLUSTER' followed by the names of the variables which contain the cluster data. Normally you will also want to specify /CRITERIA=CLUSTERS(k) where *k* is the number of clusters. If this is not given, then *k* defaults to 2.

The command uses an iterative algorithm to determine the clusters for each case. It will continue iterating until convergence, or until *max_iter* iterations have been done. The default value of *max_iter* is 2.

The MISSING subcommand determines the handling of missing variables. If INCLUDE is set, then user-missing values are considered at their face value and not as missing values. If EXCLUDE is set, which is the default, user-missing values are excluded as well as system-missing values.

If LISTWISE is set, then the entire case is excluded from the analysis whenever any of the clustering variables contains a missing value. If PAIRWISE is set, then a case is considered missing only if all the clustering variables contain missing values. Otherwise it is clustered on the basis of the non-missing values. The default is LISTWISE.

15.14 RANK

```
RANK
    [VARIABLES=] var_list [{A,D}] [BY var_list]
    /TIES={MEAN,LOW,HIGH,CONDENSE}
    /FRACTION={BLOM,TUKEY,VW,RANKIT}
    /PRINT[={YES,NO}
    /MISSING={EXCLUDE,INCLUDE}

    /RANK [INTO var_list]
    /NTILES(k) [INTO var_list]
    /NORMAL [INTO var_list]
    /PERCENT [INTO var_list]
    /RFRACTION [INTO var_list]
    /PROPORTION [INTO var_list]
    /N [INTO var_list]
```

/SAVAGE [INTO *var_list*]

The RANK command ranks variables and stores the results into new variables.

The VARIABLES subcommand, which is mandatory, specifies one or more variables whose values are to be ranked. After each variable, 'A' or 'D' may appear, indicating that the variable is to be ranked in ascending or descending order. Ascending is the default. If a BY keyword appears, it should be followed by a list of variables which are to serve as group variables. In this case, the cases are gathered into groups, and ranks calculated for each group.

The TIES subcommand specifies how tied values are to be treated. The default is to take the mean value of all the tied cases.

The FRACTION subcommand specifies how proportional ranks are to be calculated. This only has any effect if NORMAL or PROPORTIONAL rank functions are requested.

The PRINT subcommand may be used to specify that a summary of the rank variables created should appear in the output.

The function subcommands are RANK, NTILES, NORMAL, PERCENT, RFRACTION, PROPORTION and SAVAGE. Any number of function subcommands may appear. If none are given, then the default is RANK. The NTILES subcommand must take an integer specifying the number of partitions into which values should be ranked. Each subcommand may be followed by the INTO keyword and a list of variables which are the variables to be created and receive the rank scores. There may be as many variables specified as there are variables named on the VARIABLES subcommand. If fewer are specified, then the variable names are automatically created.

The MISSING subcommand determines how user missing values are to be treated. A setting of EXCLUDE means that variables whose values are user-missing are to be excluded from the rank scores. A setting of INCLUDE means they are to be included. The default is EXCLUDE.

15.15 REGRESSION

The REGRESSION procedure fits linear models to data via least-squares estimation. The procedure is appropriate for data which satisfy those assumptions typical in linear regression:

- The data set contains n observations of a dependent variable, say $Y_1, \ldots, Y_n$, and n observations of one or more explanatory variables. Let $X_{11}, X_{12}, \ldots, X_{1n}$ denote the n observations of the first explanatory variable; $X_{21}, \ldots, X_{2n}$ denote the n observations of the second explanatory variable; $X_{k1}, \ldots, X_{kn}$ denote the n observations of the kth explanatory variable.

- The dependent variable Y has the following relationship to the explanatory variables: $Y_i = b_0 + b_1 X_{1i} + \ldots + b_k X_{ki} + Z_i$ where $b_0, b_1, \ldots, b_k$ are unknown coefficients, and $Z_1, \ldots, Z_n$ are independent, normally distributed *noise* terms with mean zero and common variance. The noise, or *error* terms are unobserved. This relationship is called the *linear model*.

The REGRESSION procedure estimates the coefficients $b_0, \ldots, b_k$ and produces output relevant to inferences for the linear model.

15.15.1 Syntax

```
REGRESSION
        /VARIABLES=var_list
        /DEPENDENT=var_list
        /STATISTICS={ALL, DEFAULTS, R, COEFF, ANOVA, BCOV, CI[conf]}
        /SAVE={PRED, RESID}
```

The REGRESSION procedure reads the active dataset and outputs statistics relevant to the linear model specified by the user.

The VARIABLES subcommand, which is required, specifies the list of variables to be analyzed. Keyword VARIABLES is required. The DEPENDENT subcommand specifies the dependent variable of the linear model. The DEPENDENT subcommand is required. All variables listed in the VARIABLES subcommand, but not listed in the DEPENDENT subcommand, are treated as explanatory variables in the linear model.

All other subcommands are optional:

The STATISTICS subcommand specifies additional statistics to be displayed. The following keywords are accepted:

ALL   All of the statistics below.

R   The ratio of the sums of squares due to the model to the total sums of squares for the dependent variable.

COEFF   A table containing the estimated model coefficients and their standard errors.

CI (conf)   This item is only relevant if COEFF has also been selected. It specifies that the confidence interval for the coefficients should be printed. The optional value conf, which must be in parentheses, is the desired confidence level expressed as a percentage.

ANOVA   Analysis of variance table for the model.

BCOV   The covariance matrix for the estimated model coefficients.

DEFAULT   The same as if R, COEFF, and ANOVA had been selected.

The SAVE subcommand causes PSPP to save the residuals or predicted values from the fitted model to the active dataset. PSPP will store the residuals in a variable called 'RES1' if no such variable exists, 'RES2' if 'RES1' already exists, 'RES3' if 'RES1' and 'RES2' already exist, etc. It will choose the name of the variable for the predicted values similarly, but with 'PRED' as a prefix. When SAVE is used, PSPP ignores TEMPORARY, treating temporary transformations as permanent.

15.15.2 Examples

The following PSPP syntax will generate the default output and save the predicted values and residuals to the active dataset.

```
title 'Demonstrate REGRESSION procedure'.
data list / v0 1-2 (A) v1 v2 3-22 (10).
begin data.
b  7.735648 -23.97588
b  6.142625 -19.63854
```

```
a  7.651430  -25.26557
c  6.125125  -16.57090
a  8.245789  -25.80001
c  6.031540  -17.56743
a  9.832291  -28.35977
c  5.343832  -16.79548
a  8.838262  -29.25689
b  6.200189  -18.58219
end data.
list.
regression /variables=v0 v1 v2 /statistics defaults /dependent=v2
           /save pred resid /method=enter.
```

15.16 RELIABILITY

```
RELIABILITY
      /VARIABLES=var_list
      /SCALE (name) = {var_list, ALL}
      /MODEL={ALPHA, SPLIT[(n)]}
      /SUMMARY={TOTAL,ALL}
      /MISSING={EXCLUDE,INCLUDE}
```

The **RELIABILITY** command performs reliability analysis on the data.

The **VARIABLES** subcommand is required. It determines the set of variables upon which analysis is to be performed.

The **SCALE** subcommand determines which variables reliability is to be calculated for. If it is omitted, then analysis for all variables named in the **VARIABLES** subcommand will be used. Optionally, the *name* parameter may be specified to set a string name for the scale.

The **MODEL** subcommand determines the type of analysis. If **ALPHA** is specified, then Cronbach's Alpha is calculated for the scale. If the model is **SPLIT**, then the variables are divided into 2 subsets. An optional parameter n may be given, to specify how many variables to be in the first subset. If n is omitted, then it defaults to one half of the variables in the scale, or one half minus one if there are an odd number of variables. The default model is **ALPHA**.

By default, any cases with user missing, or system missing values for any variables given in the **VARIABLES** subcommand will be omitted from analysis. The **MISSING** subcommand determines whether user missing values are to be included or excluded in the analysis.

The **SUMMARY** subcommand determines the type of summary analysis to be performed. Currently there is only one type: **SUMMARY=TOTAL**, which displays per-item analysis tested against the totals.

15.17 ROC

```
ROC    var_list BY state_var (state_value)
       /PLOT = { CURVE [(REFERENCE)], NONE }
       /PRINT = [ SE ] [ COORDINATES ]
       /CRITERIA = [ CUTOFF({INCLUDE,EXCLUDE}) ]
```

```
      [ TESTPOS ({LARGE,SMALL}) ]
      [ CI (confidence) ]
      [ DISTRIBUTION ({FREE, NEGEXPO }) ]
      /MISSING={EXCLUDE,INCLUDE}
```

The ROC command is used to plot the receiver operating characteristic curve of a dataset, and to estimate the area under the curve. This is useful for analysing the efficacy of a variable as a predictor of a state of nature.

The mandatory *var_list* is the list of predictor variables. The variable *state_var* is the variable whose values represent the actual states, and *state_value* is the value of this variable which represents the positive state.

The optional subcommand PLOT is used to determine if and how the ROC curve is drawn. The keyword CURVE means that the ROC curve should be drawn, and the optional keyword REFERENCE, which should be enclosed in parentheses, says that the diagonal reference line should be drawn. If the keyword NONE is given, then no ROC curve is drawn. By default, the curve is drawn with no reference line.

The optional subcommand PRINT determines which additional tables should be printed. Two additional tables are available. The SE keyword says that standard error of the area under the curve should be printed as well as the area itself. In addition, a p-value under the null hypothesis that the area under the curve equals 0.5 will be printed. The COORDINATES keyword says that a table of coordinates of the ROC curve should be printed.

The CRITERIA subcommand has four optional parameters:

- The TESTPOS parameter may be LARGE or SMALL. LARGE is the default, and says that larger values in the predictor variables are to be considered positive. SMALL indicates that smaller values should be considered positive.

- The CI parameter specifies the confidence interval that should be printed. It has no effect if the SE keyword in the PRINT subcommand has not been given.

- The DISTRIBUTION parameter determines the method to be used when estimating the area under the curve. There are two possibilities, *viz*: FREE and NEGEXPO. The FREE method uses a non-parametric estimate, and the NEGEXPO method a bi-negative exponential distribution estimate. The NEGEXPO method should only be used when the number of positive actual states is equal to the number of negative actual states. The default is FREE.

- The CUTOFF parameter is for compatibility and is ignored.

The MISSING subcommand determines whether user missing values are to be included or excluded in the analysis. The default behaviour is to exclude them. Cases are excluded on a listwise basis; if any of the variables in *var_list* or if the variable *state_var* is missing, then the entire case will be excluded.

16 Utilities

Commands that don't fit any other category are placed here.

Most of these commands are not affected by commands like IF and LOOP: they take effect only once, unconditionally, at the time that they are encountered in the input.

16.1 ADD DOCUMENT

ADD DOCUMENT
'line one' 'line two' ... 'last line' .

ADD DOCUMENT adds one or more lines of descriptive commentary to the active dataset. Documents added in this way are saved to system files. They can be viewed using SYSFILE INFO or DISPLAY DOCUMENTS. They can be removed from the active dataset with DROP DOCUMENTS.

Each line of documentary text must be enclosed in quotation marks, and may not be more than 80 bytes long. See Section 16.5 [DOCUMENT], page 153.

16.2 CACHE

CACHE.

This command is accepted, for compatibility, but it has no effect.

16.3 CD

CD 'new directory' .

CD changes the current directory. The new directory will become that specified by the command.

16.4 COMMENT

Two possibles syntaxes:
COMMENT comment text ... .
*comment text ... .

COMMENT is ignored. It is used to provide information to the author and other readers of the PSPP syntax file.

COMMENT can extend over any number of lines. Don't forget to terminate it with a dot or a blank line.

16.5 DOCUMENT

DOCUMENT documentary_text.

DOCUMENT adds one or more lines of descriptive commentary to the active dataset. Documents added in this way are saved to system files. They can be viewed using SYSFILE INFO or DISPLAY DOCUMENTS. They can be removed from the active dataset with DROP DOCUMENTS.

Specify the documentary text following the DOCUMENT keyword. It is interpreted literally — any quotes or other punctuation marks will be included in the file. You can extend the documentary text over as many lines as necessary. Lines are truncated at 80 bytes. Don't forget to terminate the command with a dot or a blank line. See Section 16.1 [ADD DOCUMENT], page 153.

16.6 DISPLAY DOCUMENTS

DISPLAY DOCUMENTS.

DISPLAY DOCUMENTS displays the documents in the active dataset. Each document is preceded by a line giving the time and date that it was added. See Section 16.5 [DOCUMENT], page 153.

16.7 DISPLAY FILE LABEL

DISPLAY FILE LABEL.

DISPLAY FILE LABEL displays the file label contained in the active dataset, if any. See Section 16.12 [FILE LABEL], page 155.

This command is a PSPP extension.

16.8 DROP DOCUMENTS

DROP DOCUMENTS.

DROP DOCUMENTS removes all documents from the active dataset. New documents can be added with DOCUMENT (see Section 16.5 [DOCUMENT], page 153).

DROP DOCUMENTS changes only the active dataset. It does not modify any system files stored on disk.

16.9 ECHO

ECHO 'arbitrary text' .

Use ECHO to write arbitrary text to the output stream. The text should be enclosed in quotation marks following the normal rules for string tokens (see Section 6.1 [Tokens], page 28).

16.10 ERASE

ERASE FILE file_name.

ERASE FILE deletes a file from the local filesystem. file_name must be quoted. This command cannot be used if the SAFER (see Section 16.20 [SET], page 157) setting is active.

16.11 EXECUTE

EXECUTE.

EXECUTE causes the active dataset to be read and all pending transformations to be executed.

16.12 FILE LABEL

FILE LABEL *file_label*.

FILE LABEL provides a title for the active dataset. This title will be saved into system files and portable files that are created during this PSPP run.

file_label should not be quoted. If quotes are included, they are literally interpreted and become part of the file label.

16.13 FINISH

FINISH.

FINISH terminates the current PSPP session and returns control to the operating system.

16.14 HOST

HOST.
HOST COMMAND=['*command*'...].

HOST suspends the current PSPP session and temporarily returns control to the operating system. This command cannot be used if the SAFER (see Section 16.20 [SET], page 157) setting is active.

If the COMMAND subcommand is specified, as a sequence of shell commands as quoted strings within square brackets, then PSPP executes them together in a single subshell.

If no subcommands are specified, then PSPP invokes an interactive subshell.

16.15 INCLUDE

INCLUDE [FILE=]'*file_name*' [ENCODING='*encoding*'].

INCLUDE causes the PSPP command processor to read an additional command file as if it were included bodily in the current command file. If errors are encountered in the included file, then command processing will stop and no more commands will be processed. Include files may be nested to any depth, up to the limit of available memory.

The INSERT command (see Section 16.16 [INSERT], page 155) is a more flexible alternative to INCLUDE. An INCLUDE command acts the same as INSERT with ERROR=STOP CD=NO SYNTAX=BATCH specified.

The optional ENCODING subcommand has the same meaning as with INSERT.

16.16 INSERT

INSERT [FILE=]'*file_name*'
  [CD={NO,YES}]
  [ERROR={CONTINUE,STOP}]
  [SYNTAX={BATCH,INTERACTIVE}]
  [ENCODING={LOCALE, '*charset_name*'}].

INSERT is similar to INCLUDE (see Section 16.15 [INCLUDE], page 155) but somewhat more flexible. It causes the command processor to read a file as if it were embedded in the current command file.

If `CD=YES` is specified, then before including the file, the current directory will be changed to the directory of the included file. The default setting is 'CD=NO'. Note that this directory will remain current until it is changed explicitly (with the `CD` command, or a subsequent `INSERT` command with the 'CD=YES' option). It will not revert to its original setting even after the included file is finished processing.

If `ERROR=STOP` is specified, errors encountered in the inserted file will cause processing to immediately cease. Otherwise processing will continue at the next command. The default setting is `ERROR=CONTINUE`.

If `SYNTAX=INTERACTIVE` is specified then the syntax contained in the included file must conform to interactive syntax conventions. See Section 6.3 [Syntax Variants], page 30. The default setting is `SYNTAX=BATCH`.

`ENCODING` optionally specifies the character set used by the included file. Its argument, which is not case-sensitive, must be in one of the following forms:

LOCALE   The encoding used by the system locale, or as overridden by the `SET` command (see Section 16.20 [SET], page 157). On GNU/Linux and other Unix-like systems, environment variables, e.g. `LANG` or `LC_ALL`, determine the system locale.

charset_name
            One of the character set names listed by IANA at `http://www.iana.org/assignments/character-sets`. Some examples are `ASCII` (United States), `ISO-8859-1` (western Europe), `EUC-JP` (Japan), and `windows-1252` (Windows). Not all systems support all character sets.

Auto,encoding
            Automatically detects whether a syntax file is encoded in an Unicode encoding such as UTF-8, UTF-16, or UTF-32. If it is not, then PSPP generally assumes that the file is encoded in *encoding* (an IANA character set name). However, if *encoding* is UTF-8, and the syntax file is not valid UTF-8, PSPP instead assumes that the file is encoded in `windows-1252`.

            For best results, *encoding* should be an ASCII-compatible encoding (the most common locale encodings are all ASCII-compatible), because encodings that are not ASCII compatible cannot be automatically distinguished from UTF-8.

Auto

Auto,Locale
            Automatic detection, as above, with the default encoding taken from the system locale or the setting on `SET LOCALE`.

When ENCODING is not specified, the default is taken from the `--syntax-encoding` command option, if it was specified, and otherwise it is `Auto`.

16.17 OUTPUT

```
OUTPUT MODIFY
    /SELECT TABLES
    /TABLECELLS SELECT = [ {SIGNIFICANCE, COUNT} ]
        FORMAT = fmt_spec.
```

Please note: In the above synopsis the characters '[' and ']' are literals. They must appear in the syntax to be interpreted.

OUTPUT changes the appearance of the tables in which results are printed. In particular, it can be used to set the format and precision to which results are displayed.

After running this command, the default table appearance parameters will have been modified and each new output table generated will use the new parameters.

Following /TABLECELLS SELECT = a list of cell classes must appear, enclosed in square brackets. This list determines the classes of values should be selected for modification. Each class can be:

SIGNIFICANCE

  Significance of tests (p-values).

COUNT   Counts or sums of weights.

The value of *fmt_spec* must be a valid output format (see Section 6.7.4 [Input and Output Formats], page 34). Note that not all possible formats are meaningful for all classes.

16.18 PERMISSIONS

PERMISSIONS
    FILE='*file_name*'
    /PERMISSIONS = {READONLY,WRITEABLE}.

PERMISSIONS changes the permissions of a file. There is one mandatory subcommand which specifies the permissions to which the file should be changed. If you set a file's permission to READONLY, then the file will become unwritable either by you or anyone else on the system. If you set the permission to WRITEABLE, then the file will become writeable by you; the permissions afforded to others will be unchanged. This command cannot be used if the SAFER (see Section 16.20 [SET], page 157) setting is active.

16.19 PRESERVE and RESTORE

PRESERVE.
    . . .
RESTORE.

PRESERVE saves all of the settings that SET (see Section 16.20 [SET], page 157) can adjust. A later RESTORE command restores those settings.

PRESERVE can be nested up to five levels deep.

16.20 SET

SET

(data input)
    /BLANKS={SYSMIS,'.',number}
    /DECIMAL={DOT,COMMA}
    /FORMAT=*fmt_spec*
    /EPOCH={AUTOMATIC,*year*}

```
/RIB={NATIVE,MSBFIRST,LSBFIRST,VAX}
/RRB={NATIVE,ISL,ISB,IDL,IDB,VF,VD,VG,ZS,ZL}
```

(interaction)
```
/MXERRS=max_errs
/MXWARNS=max_warnings
/WORKSPACE=workspace_size
```

(syntax execution)
```
/LOCALE='locale'
/MEXPAND={ON,OFF}
/MITERATE=max_iterations
/MNEST=max_nest
/MPRINT={ON,OFF}
/MXLOOPS=max_loops
/SEED={RANDOM,seed_value}
/UNDEFINED={WARN,NOWARN}
/FUZZBITS=fuzzbits
```

(data output)
```
/CC{A,B,C,D,E}={'npre,pre,suf,nsuf','npre.pre.suf.nsuf'}
/DECIMAL={DOT,COMMA}
/FORMAT=fmt_spec
/WIB={NATIVE,MSBFIRST,LSBFIRST,VAX}
/WRB={NATIVE,ISL,ISB,IDL,IDB,VF,VD,VG,ZS,ZL}
```

(output routing)
```
/ERRORS={ON,OFF,TERMINAL,LISTING,BOTH,NONE}
/MESSAGES={ON,OFF,TERMINAL,LISTING,BOTH,NONE}
/PRINTBACK={ON,OFF,TERMINAL,LISTING,BOTH,NONE}
/RESULTS={ON,OFF,TERMINAL,LISTING,BOTH,NONE}
```

(output driver options)
```
/HEADERS={NO,YES,BLANK}
/LENGTH={NONE,n_lines}
/MORE={ON,OFF}
/WIDTH={NARROW,WIDTH,n_characters}
/TNUMBERS={VALUES,LABELS,BOTH}
/TVARS={NAMES,LABELS,BOTH}
```

(logging)
```
/JOURNAL={ON,OFF} ['file_name']
```

(system files)
```
/COMPRESSION={ON,OFF}
/SCOMPRESSION={ON,OFF}
```

(miscellaneous)
        /SAFER=ON
        /LOCALE='string'

(obsolete settings accepted for compatibility, but ignored)
        /BOXSTRING={'xxx','xxxxxxxxxxx'}
        /CASE={UPPER,UPLOW}
        /CPI=cpi_value
        /HIGHRES={ON,OFF}
        /HISTOGRAM='c'
        /LOWRES={AUTO,ON,OFF}
        /LPI=lpi_value
        /MENUS={STANDARD,EXTENDED}
        /MXMEMORY=max_memory
        /SCRIPTTAB='c'
        /TB1={'xxx','xxxxxxxxxxx'}
        /TBFONTS='string'
        /XSORT={YES,NO}

SET allows the user to adjust several parameters relating to PSPP's execution. Since there are many subcommands to this command, its subcommands will be examined in groups.

For subcommands that take boolean values, ON and YES are synonymous, as are OFF and NO, when used as subcommand values.

The data input subcommands affect the way that data is read from data files. The data input subcommands are

BLANKS   This is the value assigned to an item data item that is empty or contains only white space. An argument of SYSMIS or '.' will cause the system-missing value to be assigned to null items. This is the default. Any real value may be assigned.

DECIMAL

              This value may be set to DOT or COMMA. Setting it to DOT causes the decimal point character to be '.' and the grouping character to be ','. Setting it to COMMA causes the decimal point character to be ',' and the grouping character to be '.'. The default value is determined from the system locale.

FORMAT   Allows the default numeric input/output format to be specified. The default is F8.2. See Section 6.7.4 [Input and Output Formats], page 34.

EPOCH   Specifies the range of years used when a 2-digit year is read from a data file or used in a date construction expression (see Section 7.7.8.4 [Date Construction], page 54). If a 4-digit year is specified for the epoch, then 2-digit years are interpreted starting from that year, known as the epoch. If AUTOMATIC (the default) is specified, then the epoch begins 69 years before the current date.

RIB

              PSPP extension to set the byte ordering (endianness) used for reading data in IB or PIB format (see Section 6.7.4.4 [Binary and Hexadecimal Numeric Formats], page 39). In MSBFIRST ordering, the most-significant byte appears at the left end of a IB or PIB field. In LSBFIRST ordering, the least-significant byte appears at the left end. VAX ordering is like MSBFIRST, except that each pair of bytes is in reverse order. NATIVE, the default, is equivalent to MSBFIRST or LSBFIRST depending on the native format of the machine running PSPP.

RRB

PSPP extension to set the floating-point format used for reading data in RB format (see Section 6.7.4.4 [Binary and Hexadecimal Numeric Formats], page 39). The possibilities are:

NATIVE   The native format of the machine running PSPP. Equivalent to either IDL or IDB.

ISL   32-bit IEEE 754 single-precision floating point, in little-endian byte order.

ISB   32-bit IEEE 754 single-precision floating point, in big-endian byte order.

IDL   64-bit IEEE 754 double-precision floating point, in little-endian byte order.

IDB   64-bit IEEE 754 double-precision floating point, in big-endian byte order.

VF   32-bit VAX F format, in VAX-endian byte order.

VD   64-bit VAX D format, in VAX-endian byte order.

VG   64-bit VAX G format, in VAX-endian byte order.

ZS   32-bit IBM Z architecture short format hexadecimal floating point, in big-endian byte order.

ZL   64-bit IBM Z architecture long format hexadecimal floating point, in big-endian byte order.

          Z architecture also supports IEEE 754 floating point. The ZS and ZL formats are only for use with very old input files.

The default is NATIVE.

Interaction subcommands affect the way that PSPP interacts with an online user. The interaction subcommands are

MXERRS   The maximum number of errors before PSPP halts processing of the current command file. The default is 50.

MXWARNS

The maximum number of warnings + errors before PSPP halts processing the current command file. The special value of zero means that all warning situations should be ignored. No warnings will be issued, except a single initial warning advising the user that warnings will not be given. The default value is 100.

Syntax execution subcommands control the way that PSPP commands execute. The syntax execution subcommands are

LOCALE   Overrides the system locale for the purpose of reading and writing syntax and data files. The argument should be a locale name in the general form *language_country.encoding*, where *language* and *country* are 2-character language and country abbreviations, respectively, and *encoding* is an IANA character set name. Example locales are en_US.UTF-8 (UTF-8 encoded English as spoken in the United States) and ja_JP.EUC-JP (EUC-JP encoded Japanese as spoken in Japan).

MEXPAND
MITERATE
MNEST
MPRINT   Currently not used.

MXLOOPS

    The maximum number of iterations for an uncontrolled loop (see Section 14.4 [LOOP], page 125). The default *max_loops* is 40.

SEED   The initial pseudo-random number seed. Set to a real number or to RANDOM, which will obtain an initial seed from the current time of day.

UNDEFINED

    Currently not used.

FUZZBITS

    The maximum number of bits of errors in the least-significant places to accept for rounding up a value that is almost halfway between two possibilities for rounding with the RND operator (see Section 7.7.2 [Miscellaneous Mathematics], page 48). The default *fuzzbits* is 6.

WORKSPACE

    The maximum amount of memory (in kilobytes) that PSPP will use to store data being processed. If memory in excess of the workspace size is required, then PSPP will start to use temporary files to store the data. Setting a higher value will, in general, mean procedures will run faster, but may cause other applications to run slower. On platforms without virtual memory management, setting a very large workspace may cause PSPP to abort.

Data output subcommands affect the format of output data. These subcommands are

CCA
CCB
CCC
CCD
CCE

    Set up custom currency formats. See Section 6.7.4.2 [Custom Currency Formats], page 37, for details.

DECIMAL

    The default DOT setting causes the decimal point character to be '.'. A setting of COMMA causes the decimal point character to be ','.

FORMAT   Allows the default numeric input/output format to be specified. The default is
          F8.2. See Section 6.7.4 [Input and Output Formats], page 34.

WIB

          PSPP extension to set the byte ordering (endianness) used for writing data in IB
          or PIB format (see Section 6.7.4.4 [Binary and Hexadecimal Numeric Formats],
          page 39). In `MSBFIRST` ordering, the most-significant byte appears at the left
          end of a IB or PIB field. In `LSBFIRST` ordering, the least-significant byte appears
          at the left end. `VAX` ordering is like `MSBFIRST`, except that each pair of bytes
          is in reverse order. `NATIVE`, the default, is equivalent to `MSBFIRST` or `LSBFIRST`
          depending on the native format of the machine running PSPP.

WRB

          PSPP extension to set the floating-point format used for writing data in RB for-
          mat (see Section 6.7.4.4 [Binary and Hexadecimal Numeric Formats], page 39).
          The choices are the same as `SET RIB`. The default is `NATIVE`.

In the PSPP text-based interface, the output routing subcommands affect where output
is sent. The following values are allowed for each of these subcommands:

OFF

NONE   Discard this kind of output.

TERMINAL
          Write this output to the terminal, but not to listing files and other output
          devices.

LISTING   Write this output to listing files and other output devices, but not to the ter-
          minal.

ON
BOTH   Write this type of output to all output devices.

These output routing subcommands are:

ERRORS   Applies to error and warning messages. The default is `BOTH`.

MESSAGES
          Applies to notes. The default is `BOTH`.

PRINTBACK
          Determines whether the syntax used for input is printed back as part of the
          output. The default is `NONE`.

RESULTS   Applies to everything not in one of the above categories, such as the results of
          statistical procedures. The default is `BOTH`.

These subcommands have no effect on output in the PSPP GUI environment.

Output driver option subcommands affect output drivers' settings. These subcommands
are

HEADERS
LENGTH
MORE
WIDTH
TNUMBERS

>The TNUMBERS option sets the way in which values are displayed in output tables. The valid settings are VALUES, LABELS and BOTH. If TNUMBERS is set to VALUES, then all values are displayed with their literal value (which for a numeric value is a number and for a string value an alphanumeric string). If TNUMBERS is set to LABELS, then values are displayed using their assigned labels if any. (See Section 11.12 [VALUE LABELS], page 105.) If the a value has no label, then it will be displayed using its literal value. If TNUMBERS is set to BOTH, then values will be displayed with both their label (if any) and their literal value in parentheses.

TVARS   The TVARS option sets the way in which variables are displayed in output tables. The valid settings are NAMES, LABELS and BOTH. If TVARS is set to NAMES, then all variables are displayed using their names. If TVARS is set to LABELS, then variables are displayed using their label if one has been set. If no label has been set, then the name will be used. (See Section 11.15 [VARIABLE LABELS], page 107.) If TVARS is set to BOTH, then variables will be displayed with both their label (if any) and their name in parentheses.

Logging subcommands affect logging of commands executed to external files. These subcommands are

JOURNAL
LOG   These subcommands, which are synonyms, control the journal. The default is ON, which causes commands entered interactively to be written to the journal file. Commands included from syntax files that are included interactively and error messages printed by PSPP are also written to the journal file, prefixed by '>'. OFF disables use of the journal.

>The journal is named pspp.jnl by default. A different name may be specified.

System file subcommands affect the default format of system files produced by PSPP. These subcommands are

COMPRESSION

>Not currently used.

SCOMPRESSION

>Whether system files created by SAVE or XSAVE are compressed by default. The default is ON.

Security subcommands affect the operations that commands are allowed to perform. The security subcommands are

SAFER   Setting this option disables the following operations:

>• The ERASE command.

>• The HOST command.

>• The PERMISSIONS command.

- Pipes (file names beginning or ending with '|').

Be aware that this setting does not guarantee safety (commands can still overwrite files, for instance) but it is an improvement. When set, this setting cannot be reset during the same session, for obvious security reasons.

LOCALE This item is used to set the default character encoding. The encoding may be specified either as an encoding name or alias (see http://www.iana.org/assignments/character-sets), or as a locale name. If given as a locale name, only the character encoding of the locale is relevant.

System files written by PSPP will use this encoding. System files read by PSPP, for which the encoding is unknown, will be interpreted using this encoding.

The full list of valid encodings and locale names/alias are operating system dependent. The following are all examples of acceptable syntax on common GNU/Linux systems.

```
SET LOCALE='iso-8859-1'.

SET LOCALE='ru_RU.cp1251'.

SET LOCALE='japanese'.
```

Contrary to the intuition, this command does not affect any aspect of the system's locale.

16.21 SHOW

```
SHOW
        [ALL]
        [BLANKS]
        [CC]
        [CCA]
        [CCB]
        [CCC]
        [CCD]
        [CCE]
        [COPYING]
        [DECIMALS]
        [DIRECTORY]
        [ENVIRONMENT]
        [FORMAT]
        [FUZZBITS]
        [LENGTH]
        [MXERRS]
        [MXLOOPS]
        [MXWARNS]
        [N]
        [SCOMPRESSION]
        [TEMPDIR]
        [UNDEFINED]
```

[VERSION]
[WARRANTY]
[WEIGHT]
[WIDTH]

SHOW can be used to display the current state of PSPP's execution parameters. Parameters that can be changed using SET (see Section 16.20 [SET], page 157), can be examined using SHOW using the subcommand with the same name. SHOW supports the following additional subcommands:

ALL   Show all settings.

CC   Show all custom currency settings (CCA through CCE).

DIRECTORY
           Shows the current working directory.

ENVIRONMENT
           Shows the operating system details.

N   Reports the number of cases in the active dataset. The reported number is not weighted. If no dataset is defined, then 'Unknown' will be reported.

TEMPDIR   Shows the path of the directory where temporary files will be stored.

VERSION   Shows the version of this installation of PSPP.

WARRANTY   Show details of the lack of warranty for PSPP.

COPYING / LICENSE
           Display the terms of PSPP's copyright licence (see Chapter 2 [License], page 3).

Specifying SHOW without any subcommands is equivalent to SHOW ALL.

16.22 SUBTITLE

> SUBTITLE 'subtitle_string'.
>   or
> SUBTITLE subtitle_string.

SUBTITLE provides a subtitle to a particular PSPP run. This subtitle appears at the top of each output page below the title, if headers are enabled on the output device.

Specify a subtitle as a string in quotes. The alternate syntax that did not require quotes is now obsolete. If it is used then the subtitle is converted to all uppercase.

16.23 TITLE

> TITLE 'title_string'.
>   or
> TITLE title_string.

TITLE provides a title to a particular PSPP run. This title appears at the top of each output page, if headers are enabled on the output device.

Specify a title as a string in quotes. The alternate syntax that did not require quotes is now obsolete. If it is used then the title is converted to all uppercase.

17 Invoking `pspp-convert`

`pspp-convert` is a command-line utility accompanying PSPP. It reads an SPSS system or portable file *input* and writes a copy of it to another *output* in a different format. Synopsis:

    `pspp-convert` [*options*] *input output*

    `pspp-convert --help`

    `pspp-convert --version`

The format of *input* is automatically detected, when possible. The character encoding of old SPSS system files cannot always be guessed correctly, and SPSS/PC+ system files do not include any indication of their encoding. Use `-e` *encoding* to specify the encoding in this case.

By default, the intended format for *output* is inferred based on its extension:

`csv`

`txt`   Comma-separated value. Each value is formatted according to its variable's print format. The first line in the file contains variable names.

`sav`

`sys`   SPSS system file.

`por`   SPSS portable file.

As a special case of format conversion, `pspp-convert` can decrypt an encrypted SPSS system file. Specify the encrypted file as *input*. The output will be the equivalent plaintext SPSS system file. You will be prompted for the password (or use `-p`, documented below).

Use `-O` *extension* to override the inferred format or to specify the format for unrecognized extensions.

The following options are accepted:

`-O` *format*
`--output-format=`*format*
> Specifies the desired output format. *format* must be one of the extensions listed above, e.g. `-O csv` requests comma-separated value output.

`-c` *maxcases*
`--cases=`*maxcases*
> By default, all cases are copied from *input* to *output*. Specifying this option to limit the number of cases written to *output* to *maxcases*.

`-e` *charset*
`--encoding=`*charset*
> Overrides the encoding in which character strings in *input* are interpreted. This option is necessary because old SPSS system files, and SPSS/PC+ system files, do not self-identify their encoding.

`-p` *password*
`--password=`*password*

> Specifies the password to use to decrypt an encrypted SPSS system file. If this option is not specified, `pspp-convert` will prompt interactively for the password as necessary.
>
> Be aware that command-line options, including passwords, may be visible to other users on multiuser systems.

`-h`
`--help`   Prints a usage message on stdout and exits.

`-v`
`--version`

> Prints version information on stdout and exits.

18 Invoking `pspp-dump-sav`

`pspp-dump-sav` is a command-line utility accompanying PSPP. It reads one or more SPSS system files and prints their contents. The output format is useful for debugging system file readers and writers and for discovering how to interpret unknown or poorly understood records. End users may find the output useful for providing the PSPP developers information about system files that PSPP does not accurately read.

Synopsis:

> `pspp-dump-sav` [-d[*maxcases*] | --data[=*maxcases*]] *file*...

> `pspp-dump-sav` --help | -h

> `pspp-dump-sav` --version | -v

The following options are accepted:

-d[*maxcases*]
--data[=*maxcases*]

> By default, `pspp-dump-sav` does not print any of the data in a system file, only the file headers. Specify this option to print the data as well. If *maxcases* is specified, then it limits the number of cases printed.

-h
--help   Prints a usage message on stdout and exits.

-v
--version

> Prints version information on stdout and exits.

Some errors that prevent files from being interpreted successfully cause `pspp-dump-sav` to exit without reading any additional files given on the command line.

19 Not Implemented

This chapter lists parts of the PSPP language that are not yet implemented.

2SLS   Two stage least squares regression

ACF   Autocorrelation function

ALSCAL   Multidimensional scaling

ANACOR   Correspondence analysis

ANOVA   Factorial analysis of variance

CASEPLOT   Plot time series

CASESTOVARS
            Restructure complex data

CATPCA   Categorical principle components analysis

CATREG   Categorical regression

CCF   Time series cross correlation

CLEAR TRANSFORMATIONS
            Clears transformations from active dataset

CLUSTER   Hierarchical clustering

CONJOINT   Analyse full concept data

CORRESPONDENCE
            Show correspondence

COXREG   Cox proportional hazards regression

CREATE   Create time series data

CSDESCRIPTIVES
            Complex samples descriptives

CSGLM   Complex samples GLM

CSLOGISTIC
            Complex samples logistic regression

CSPLAN   Complex samples design

CSSELECT   Select complex samples

CSTABULATE
            Tabulate complex samples

CTABLES   Display complex samples

CURVEFIT   Fit curve to line plot

DATE   Create time series data

DEFINE   Syntax macros

DETECTANOMALY
          Find unusual cases

DISCRIMINANT
          Linear discriminant analysis

EDIT   obsolete

END FILE TYPE
          Ends complex data input

FILE TYPE   Complex data input

FIT   Goodness of Fit

GENLOG   Categorical model fitting

GET TRANSLATE
          Read other file formats

GGRAPH   Custom defined graphs

HILOGLINEAR
          Hierarchical loglinear models

HOMALS   Homogeneity analysis

IGRAPH   Interactive graphs

INFO   Local Documentation

KEYED DATA LIST
          Read nonsequential data

KM   Kaplan-Meier

LOGLINEAR
          General model fitting

MANOVA   Multivariate analysis of variance

MAPS   Geographical display

MATRIX   Matrix processing

MATRIX DATA
          Matrix data input

MCONVERT   Convert covariance/correlation matrices

MIXED   Mixed linear models

MODEL CLOSE
          Close server connection

MODEL HANDLE
          Define server connection

MODEL LIST
          Show existing models

MODEL NAME
              Specify model label

MULTIPLE CORRESPONDENCE
              Multiple correspondence analysis

MULT RESPONSE
              Multiple response analysis

MVA   Missing value analysis

NAIVEBAYES
              Small sample bayesian prediction

NLR   Non Linear Regression

NOMREG   Multinomial logistic regression

NONPAR CORR
              Nonparametric correlation

NUMBERED

OLAP CUBES
              On-line analytical processing

OMS   Output management

ORTHOPLAN
              Orthogonal effects design

OVERALS   Nonlinear canonical correlation

PACF   Partial autocorrelation

PARTIAL CORR
              Partial correlation

PLANCARDS
              Conjoint analysis planning

PLUM   Estimate ordinal regression models

POINT   Marker in keyed file

PPLOT   Plot time series variables

PREDICT   Specify forecast period

PREFSCAL   Multidimensional unfolding

PRINCALS   PCA by alternating least squares

PROBIT   Probit analysis

PROCEDURE OUTPUT
              Specify output file

PROXIMITIES
              Pairwise similarity

PROXSCAL   Multidimensional scaling of proximity data

RATIO STATISTICS
            Descriptives of ratios

READ MODEL
            Read new model

RECORD TYPE
            Defines a type of record within FILE TYPE

REFORMAT   Read obsolete files

REPEATING DATA
            Specify multiple cases per input record

REPORT   Pretty print working file

RMV   Replace missing values

SCRIPT   Run script file

SEASON   Estimate seasonal factors

SELECTPRED
            Select predictor variables

SPCHART   Plot control charts

SPECTRA   Plot spectral density

STEMLEAF   Plot stem-and-leaf display

SUMMARIZE
            Univariate statistics

SURVIVAL   Survival analysis

TDISPLAY   Display active models

TREE   Create classification tree

TSAPPLY   Apply time series model

TSET   Set time sequence variables

TSHOW   Show time sequence variables

TSMODEL   Estimate time series model

TSPLOT   Plot time sequence variables

TWOSTEP CLUSTER
            Cluster observations

UNIANOVA   Univariate analysis

UNNUMBERED
            obsolete

VALIDATEDATA
            Identify suspicious cases

VARCOMP   Estimate variance

VARSTOCASES
             Restructure complex data

VERIFY   Report time series

WLS   Weighted least squares regression

XGRAPH   High resolution charts

20 Bugs

Occasionally you may encounter a bug in PSPP.

20.1 When to report bugs

If you discover a bug, please first:

- Make sure that it really is a bug. Sometimes, what may appear to be a bug, turns out to be a misunderstanding of how to use the program. If you are unsure, ask for advice on the pspp-users mailing list. Information about the mailing list is at `http://lists.gnu.org/mailman/listinfo/pspp-users`.

- Try an up to date version of PSPP; the problem may have been recently fixed.

- If the problem persists in the up to date version, check to see if it has already been reported. Reported issues are listed at `http://savannah.gnu.org/bugs/?group=pspp`. For known issues in individual language features, see the relevant section in see Chapter 6 [Language], page 28.

- If the problem exists in a recent version and it has not already been reported, then please report it.

20.2 How to report bugs

The best way to send a bug report is using the web page at `http://savannah.gnu.org/bugs/?group=pspp`. Alternatively, bug reports may be sent by email to `bug-gnu-pspp@gnu.org`.

In your bug report please include:

- The version of PSPP in which you encountered the problem. That means the precise version number. Do not simply say "the latest version" — releases happen quickly, and bug reports are archived indefinitely.

- The operating system and type of computer on which it is running. On a GNU or other unix-like system, the output from the `uname` command is helpful.

- A sample of the syntax which causes the problem or, if it is a user interface problem, the sequence of steps required to reproduce it. Screen shots are not usually helpful unless you are reporting a bug in the graphical user interface itself.

- A description of what you think is wrong: What happened that you didn't expect, and what did you expect to happen?

The following is an example of a useful bug report:

```
When I run PSPP 0.8.4 on the system:
"Linux knut 3.5.3-gnu #1 PREEMPT Tue Aug 28 10:49:41 UTC 2012 mips64 GNU/Linux"
Executing the following syntax:

DATA LIST FREE /x *.
BEGIN DATA.
1 2 3
END DATA.
LIST.
```

```
results in:
```

```
4
5
6
```

```
I think the output should be:
```

```
1
2
3
```

Here, the developers have the necessary information to reproduce the circumstances of the bug report, and they understand what the reporter expected.

Conversely, the following is a useless bug report:

```
I downloaded the latest version of PSPP and entered a sequence of numbers,
but when I analyse them it gives the wrong output.
```

In that example, it is impossible to reproduce, and there is no indication of why the reporter thought what he saw was wrong.

Note that the purpose of bug reports is to help improve the quality of PSPP for the benefit of all users. It is not a consultancy or support service. If that is what you want, you are welcome to make private arrangements. Since PSPP is free software, consultants have access to the information they need to provide such support. The PSPP developers appreciate all users' feedback, but cannot promise an immediate response.

Please do not use the bug reporting address for general enquiries or to seek help in using, installing or running the program. For that, use the pspp-users mailing list mentioned above.

21 Function Index

(

`(variable)` 57

A

`ABS` 48
`ACOS` 49
`ANY` 50
`ARCOS` 49
`ARSIN` 49
`ARTAN` 49
`ASIN` 49
`ATAN` 49

C

`CDF.BERNOULLI` 62
`CDF.BETA` 58
`CDF.BINOM` 62
`CDF.CAUCHY` 59
`CDF.CHISQ` 59
`CDF.EXP` 59
`CDF.F` 59
`CDF.GAMMA` 59
`CDF.GEOM` 62
`CDF.HYPER` 62
`CDF.LAPLACE` 60
`CDF.LNORMAL` 60
`CDF.LOGISTIC` 60
`CDF.NEGBIN` 62
`CDF.NORMAL` 60
`CDF.PARETO` 61
`CDF.POISSON` 62
`CDF.RAYLEIGH` 61
`CDF.T` 61
`CDF.T1G` 61
`CDF.T2G` 61
`CDF.UNIFORM` 61
`CDF.VBNOR` 58
`CDF.WEIBULL` 61
`CDFNORM` 60
`CFVAR` 50
`CONCAT` 51
`COS` 49
`CTIME.DAYS` 53
`CTIME.HOURS` 54
`CTIME.MINUTES` 54
`CTIME.SECONDS` 54

D

`DATE.DMY` 54
`DATE.MDY` 54
`DATE.MOYR` 54

`DATE.QYR` 54
`DATE.WKYR` 54
`DATE.YRDAY` 55
`DATEDIFF` 56
`DATESUM` 56

E

`EXP` 48

I

`IDF.BETA` 58
`IDF.CAUCHY` 59
`IDF.CHISQ` 59
`IDF.EXP` 59
`IDF.F` 59
`IDF.GAMMA` 59
`IDF.LAPLACE` 60
`IDF.LNORMAL` 60
`IDF.LOGISTIC` 60
`IDF.NORMAL` 60
`IDF.PARETO` 61
`IDF.RAYLEIGH` 61
`IDF.T` 61
`IDF.T1G` 61
`IDF.T2G` 61
`IDF.UNIFORM` 61
`IDF.WEIBULL` 61
`INDEX` 51

L

`LAG` 57
`LENGTH` 51
`LG10` 48
`LN` 48
`LNGAMMA` 48
`LOWER` 51
`LPAD` 51
`LTRIM` 51, 52

M

`MAX` 50
`MEAN` 50
`MIN` 50
`MISSING` 49
`MOD` 48
`MOD10` 48

N

`NCDF.BETA` 58
`NCDF.CHISQ` 59

NMISS 49
NORMAL 60
NPDF.BETA 58
NUMBER 52
NVALID 49

P

PDF.BERNOULLI 62
PDF.BETA 58
PDF.BINOM 62
PDF.BVNOR 58
PDF.CAUCHY 59
PDF.EXP 59
PDF.F 59
PDF.GAMMA 59
PDF.GEOM 62
PDF.HYPER 62
PDF.LANDAU 59
PDF.LAPLACE 60
PDF.LNORMAL 60
PDF.LOG 62
PDF.LOGISTIC 60
PDF.NEGBIN 62
PDF.NORMAL 60
PDF.NTAIL 60
PDF.PARETO 61
PDF.POISSON 62
PDF.RAYLEIGH 61
PDF.RTAIL 61
PDF.T 61
PDF.T1G 61
PDF.T2G 61
PDF.UNIFORM 61
PDF.WEIBULL 61
PDF.XPOWER 59
PROBIT 60

R

RANGE 50
RINDEX 52
RND 48
RPAD 52
RTRIM 52
RV.BERNOULLI 62
RV.BETA 58
RV.BINOM 62
RV.CAUCHY 59
RV.CHISQ 59
RV.EXP 59
RV.F 59
RV.GAMMA 59
RV.GEOM 62
RV.HYPER 62
RV.LANDAU 59
RV.LAPLACE 60
RV.LEVY 60

RV.LNORMAL 60
RV.LOG 62
RV.LOGISTIC 60
RV.LVSKEW 60
RV.NEGBIN 62
RV.NORMAL 60
RV.NTAIL 60
RV.PARETO 61
RV.POISSON 62
RV.RAYLEIGH 61
RV.RTAIL 61
RV.T 61
RV.UNIFORM 61
RV.WEIBULL 62
RV.XPOWER 59

S

SD 51
SIG.CHISQ 59
SIG.F 59
SIN 49
SQRT 48
STRING 52
SUBSTR 52, 53
SUM 51
SYSMIS 49

T

TAN 49
TIME.DAYS 53
TIME.HMS 53
TRUNC 49

U

UNIFORM 61
UPCASE 53

V

VALUE 50
VARIANCE 51

X

XDATE.DATE 55
XDATE.HOUR 55
XDATE.JDAY 55
XDATE.MDAY 55
XDATE.MINUTE 55
XDATE.MONTH 55
XDATE.QUARTER 55
XDATE.SECOND 55
XDATE.TDAY 55
XDATE.TIME 56
XDATE.WEEK 56

XDATE.WKDAY 56
XDATE.YEAR................................... 56

Y

YRMODA 57

22 Command Index

*

* 153

A

ADD DOCUMENT 153
ADD FILES 97
ADD VALUE LABELS 100
AGGREGATE 110
APPLY DICTIONARY 81
AUTORECODE 113

B

BEGIN DATA 64
BINOMIAL 141
BREAK 124

C

CACHE 153
CD 153
CHISQUARE 142
Cochran 142
COMMENT 153
COMPUTE 113
CORRELATIONS 132
COUNT 114
CROSSTABS 133

D

DATA LIST 66
DATA LIST FIXED 66
DATA LIST FREE 69
DATA LIST LIST 70
DATAFILE ATTRIBUTE 64
DATASET 65
DATASET ACTIVATE 65
DATASET CLOSE 65
DATASET COPY 65
DATASET DECLARE 65
DATASET DISPLAY 66
DATASET NAME 65
DELETE VARIABLES 100
DESCRIPTIVES 127
DISPLAY 100
DISPLAY DOCUMENTS 154
DISPLAY FILE LABEL 154
DO IF 124
DO REPEAT 124
DOCUMENT 153
DROP DOCUMENTS 154

E

ECHO 154
END CASE 70
END DATA 64
END FILE 70
ERASE 154
EXAMINE 130
EXECUTE 154
EXPORT 82

F

FACTOR 136
FILE HANDLE 70
FILE LABEL 155
FILTER 120
FINISH 155
FLIP 115
FORMATS 101
FREQUENCIES 128
FRIEDMAN 142

G

GET 15, 82
GET DATA 83
GRAPH 132

H

HOST 155

I

IF 116
IMPORT 89
INCLUDE 155
INPUT PROGRAM 73
INSERT 155

K

K-S 143
K-W 143
KENDALL 143
KOLMOGOROV-SMIRNOV 143
KRUSKAL-WALLIS 143

L

LEAVE 101
LIST 15, 76
LOGISTIC REGRESSION 138
LOOP 125

M

M-W 144
MANN-WHITNEY 144
MATCH FILES 98
MCNEMAR 144
MEANS 139
MEDIAN 144
MISSING VALUES 102
MODIFY VARS 103
MRSETS 103

N

N OF CASES 120
NEW FILE 76
NPAR TESTS 141
NUMERIC 105

O

ONEWAY 147
OUTPUT 156

P

PERMISSIONS 157
PRESERVE 157
PRINT 76
PRINT EJECT 77
PRINT FORMATS 105
PRINT SPACE 78

Q

QUICK CLUSTER 148

R

RANK 148
RECODE 116
REGRESSION 24, 150
RELIABILITY 151
RENAME VARIABLES 105
REPEATING DATA 78
REREAD 78
RESTORE 157
ROC 151
RUNS 144

S

SAMPLE 121
SAVE 16, 89
SAVE TRANSLATE 91
SELECT IF 121
SET 157
SHOW 164
SIGN 145
SORT CASES 119
SPLIT FILE 121
STRING 106
SUBTITLE 165
SYSFILE INFO 93

T

T-TEST 23, 145
TEMPORARY 122
TITLE 165

U

UPDATE 99

V

VALUE LABELS 105
VARIABLE ALIGNMENT 107
VARIABLE ATTRIBUTE 106
VARIABLE LABELS 107
VARIABLE LEVEL 108
VARIABLE ROLE 108
VARIABLE WIDTH 108
VECTOR 109

W

WEIGHT 123
WILCOXON 145
WRITE 80
WRITE FORMATS 109

X

XEXPORT 93
XSAVE 94

23 Concept Index

PSPP language 2

"

'"' 28

$

$CASENUM 34
$DATE 34
$JDATE 34
$LENGTH 34
$SYSMIS 34
$TIME 34
$WIDTH 34

&

'&' 47

,

',' 28

(

(.... 48
'( )' 46

)

) 48

*

'*' 46
'**' 47

+

'+' 46

,

PSPP, command structure 29
PSPP, invoking 4
PSPP, language 28

-

'-' 46, 47

.

'.' 32
.... 45

/

'/' 46

<

< 47
<= 47
<> 47

=

'=' 47

>

'>' 47
>= 47

_

'_' 32

'

"is defined as" 45

|

'|' 47

~

'~' 47
~= 47

A

absolute value 48
addition 46
analysis of variance 147
AND 47
ANOVA 147
arccosine 49
arcsine 49
arctangent 49
Area under curve 151
arguments, invalid 54
arguments, minimum valid 50
arguments, of date construction functions 54

arguments, of date extraction functions 55
arithmetic mean............................. 140
arithmetic operators......................... 46
attributes of variables....................... 32

B

Backus-Naur Form 45
bar chart 129, 136
Batch syntax 30
binary formats............................... 39
binomial test 141
bivariate logistic regression 138
BNF 45
Boolean................................. 46, 47
boxplot 130
bugs 174

C

case conversion 53
case-sensitivity 28
cases 64
changing directory.......................... 153
changing file permissions.................... 157
chi-square................................. 135
chisquare.................................. 135
chisquare test 142
clustering 148
Cochran Q test............................. 142
coefficient of concordance................... 143
coefficient of variation 50
comma separated values 16
command file............................... 44
command syntax, description of............... 45
commands, ordering 31
commands, structure 29
commands, unimplemented 169
concatenation 51
conditionals 124
consistency 19
constructing dates 54
constructing times........................... 53
control flow 124
convention, TO 34
copyright.................................... 3
correlation 133
cosine...................................... 49
covariance................................. 133
Cronbach's Alpha 151
cross-case function.......................... 57
currency formats 37
custom attributes........................... 33

D

data.. 64
data file 44

data files 85
data reduction.............................. 136
data, embedding in syntax files 66
Data, embedding in syntax files............... 64
data, fixed-format, reading................... 66
data, reading from a file 66
databases............................... 16, 84
dataset.................................... 32
date examination 55
date formats................................ 40
date, Julian 57
dates...................................... 53
dates, concepts 53
dates, constructing 54
dates, day of the month...................... 55
dates, day of the week 56
dates, day of the year....................... 55
dates, day-month-year 54
dates, in days............................... 55
dates, in hours.............................. 55
dates, in minutes 55
dates, in months............................ 55
dates, in quarters........................... 55
dates, in seconds............................ 55
dates, in weekdays.......................... 56
dates, in weeks 56
dates, in years.............................. 56
dates, mathematical properties of 56
dates, month-year 54
dates, quarter-year 54
dates, time of day 56
dates, valid................................ 53
dates, week-year 54
dates, year-day 55
day of the month 55
day of the week............................. 56
day of the year 55
day-month-year............................ 54
days..................................... 53, 55
decimal places.............................. 156
description of command syntax 45
deviation, standard.......................... 51
dictionary.................................. 32
directory................................... 153
division 46
DocBook 2

E

embedding data in syntax files................ 66
Embedding data in syntax files 64
embedding fixed-format data.................. 66
encoding, characters......................... 164
EQ.. 47
equality, testing 47
erroneous data.............................. 16
errors, in data.............................. 16
examination, of times........................ 53

Exploratory data analysis............... 130, 132
exponentiation 47
expression.................................... 45
expressions, mathematical 46
extraction, of dates.......................... 55
extraction, of time.......................... 53

F

factor analysis............................... 136
false... 47
file definition commands..................... 30
file handles................................... 44
file mode..................................... 157
file, command 44
file, data 44
file, output 44
file, portable................................. 44
file, syntax file.............................. 44
file, system 44
fixed-format data, reading 66
flow of control.............................. 124
formats....................................... 34
Friedman test 142
function, cross-case.......................... 57
functions 48
functions, miscellaneous 57
functions, missing-value...................... 49
functions, statistical 50
functions, string 51
functions, time & date....................... 53

G

GE... 47
geometric mean 140
Gnumeric 84
Graphic user interface 12
greater than.................................. 47
greater than or equal to 47
grouping operators 46
GT... 47

H

harmonic mean................................ 140
headers....................................... 163
hexadecimal formats.......................... 39
histogram.......................... 129, 130, 132
hours..................................... 54, 55
hours-minutes-seconds......................... 53
HTML....................................... 2, 9
Hypothesis testing............................ 23

I

identifiers 28
identifiers, reserved 28

inequality, testing............................ 47
input... 64
input program commands...................... 30
integer....................................... 45
integers 28
Interactive syntax 30
intersection, logical.......................... 47
introduction 2
inverse cosine................................ 49
inverse sine.................................. 49
inverse tangent 49
inversion, logical............................. 47
Inverting data............................... 19
invocation 4
Invocation............................. 166, 168

J

Julian date................................... 57

K

K-means clustering........................... 148
Kendall's W test 143
keywords..................................... 45
Kolmogorov-Smirnov test.................... 143
Kruskal-Wallis test 143

L

labels, value................................. 33
labels, variable 33
language, PSPP.......................... 2, 28
language, command structure................. 29
language, lexical analysis 28
language, tokens.............................. 28
LE... 47
length.. 163
less than 47
less than or equal to 47
lexical analysis 28
licence.. 3
license.. 3
Likert scale................................... 19
linear regression 24, 149
locale.. 164
logarithms.................................... 48
logical intersection........................... 47
logical inversion 47
logical operators 47
logical union................................. 47
logistic regression........................... 138
loops... 124
LT... 47

M

Mann-Whitney U test 144

mathematical expressions 46
mathematics.................................... 48
mathematics, advanced 48
mathematics, applied to times & dates 56
mathematics, miscellaneous 48
maximum 50
McNemar test 144
mean 50
means... 139
Median test 144
membership, of set 50
memory, amount used to store cases 161
minimum....................................... 50
minimum valid number of arguments.......... 50
minutes.................................... 54, 55
missing values 32, 33, 49
mode.. 157
modulus....................................... 48
modulus, by 10............................... 48
month-year.................................... 54
months 55
more.. 163
multiplication 46

N

names, of functions 48
NE... 47
negation...................................... 47
nonparametric tests 141
nonterminals 45
normality, testing 20, 130, 132
NOT... 47
npplot.. 130
null hypothesis 23
number 45
numbers....................................... 28
numbers, converting from strings.............. 52
numbers, converting to strings 52
numeric formats 35

O

obligations, your............................... 3
observations 64
OpenDocument.............................. 84
operations, order of.......................... 63
operator precedence 63
operators.......................... 29, 45, 48
operators, arithmetic 46
operators, grouping........................... 46
operators, logical 47
OR.. 47
order of commands 31
order of operations 63
output 64
output file.................................... 44

P

p-value 23
padding strings............................... 52
pager... 163
parentheses 46, 48
PDF 2, 7
percentiles............................. 129, 131
period.. 32
piechart 129
portable file 44
postgres...................................... 84
Postscript 7
PostScript 2
precedence, operator.......................... 63
precision, of output.......................... 156
principal axis factoring 136
principal components analysis 136
print format.................................. 33
procedures 31
productions 45
pspp-convert 166
pspp-dump-sav 168
PSPPIRE 12
punctuators............................. 29, 45

Q

Q, Cochran Q 142
quarter-year 54
quarters...................................... 55

R

reading data.................................. 15
reading data from a file...................... 66
reading fixed-format data 66
reals.. 28
Receiver Operating Characteristic 151
recoding data................................ 18
regression 149
reliability.................................... 19
reserved identifiers........................... 28
restricted transformations.................... 30
rights, your................................... 3
rounding 48
runs test 144

S

saving.. 16
scatterplot 132
scratch variables............................. 43
screening 16
searching strings............................. 51
seconds................................... 54, 55
set membership.............................. 50
sign test.................................... 145
sine 49

spreadlevel plot............................ 130
spreadsheet files........................... 84
spreadsheets................................ 16
square roots................................ 48
standard deviation.......................... 51
start symbol................................ 45
statistics.................................. 50
string...................................... 45
string formats.............................. 43
string functions............................ 51
strings..................................... 28
strings, case of........................ 51, 53
strings, concatenation of................... 51
strings, converting from numbers............ 52
strings, converting to numbers.............. 52
strings, finding length of.................. 51
strings, padding........................ 51, 52
strings, searching backwards................ 52
strings, taking substrings of............... 52
strings, trimming....................... 51, 52
substrings.................................. 52
subtraction................................. 46
sum... 51
SVG... 7
symbol, start............................... 45
syntax file................................. 44
SYSMIS...................................... 18
system file................................. 44
system files................................ 15
system variables............................ 34
system-missing.............................. 47

T

T-test...................................... 23
tangent..................................... 49
terminals................................... 45
terminals and nonterminals, differences..... 45
testing for equality........................ 47
testing for inequality...................... 47
text files.................................. 85
time.. 56
time examination............................ 53
time formats................................ 40
time, concepts.............................. 53
time, in days........................... 53, 55
time, in hours.......................... 54, 55
time, in hours-minutes-seconds.............. 53
time, in minutes........................ 54, 55
time, in seconds........................ 54, 55
time, instants of........................... 53
time, intervals............................. 53
time, lengths of............................ 53
time, mathematical properties of........... 56
times....................................... 53
times, constructing......................... 53
times, in days.............................. 55

tnumbers.................................... 163
TO convention............................... 34
tokens...................................... 28
transformation.............................. 16
transformations........................ 30, 110
trigonometry................................ 49
troubleshooting............................. 174
true.. 47
truncation.................................. 49
type of variables........................... 33

U

U, Mann-Whitney U........................... 144
unimplemented commands...................... 169
union, logical.............................. 47
utility commands............................ 30

V

value label................................. 57
value labels................................ 33
values, Boolean............................. 46
values, missing..................... 32, 33, 49
values, system-missing...................... 47
var-list.................................... 45
var-name.................................... 45
variable.................................... 32
variable labels............................. 33
variable names, ending with period.......... 32
variable role............................... 33
variables................................... 14
variables, attributes of.................... 32
variables, system........................... 34
variables, type............................. 33
variables, width............................ 33
variance.................................... 51
variation, coefficient of................... 50

W

week.. 56
week-year................................... 54
weekday..................................... 56
white space, trimming................... 51, 52
width....................................... 163
width of variables.......................... 33
wilcoxon matched pairs signed ranks test.... 145
workspace................................... 161
write format................................ 33

Y

year-day.................................... 55
years....................................... 56
your rights and obligations................. 3

Appendix A GNU Free Documentation License

Version 1.3, 3 November 2008

Copyright © 2000, 2001, 2002, 2007, 2008 Free Software Foundation, Inc.
`http://fsf.org/`

0. PREAMBLE

The purpose of this License is to make a manual, textbook, or other functional and useful document *free* in the sense of freedom: to assure everyone the effective freedom to copy and redistribute it, with or without modifying it, either commercially or non-commercially. Secondarily, this License preserves for the author and publisher a way to get credit for their work, while not being considered responsible for modifications made by others.

This License is a kind of "copyleft", which means that derivative works of the document must themselves be free in the same sense. It complements the GNU General Public License, which is a copyleft license designed for free software.

We have designed this License in order to use it for manuals for free software, because free software needs free documentation: a free program should come with manuals providing the same freedoms that the software does. But this License is not limited to software manuals; it can be used for any textual work, regardless of subject matter or whether it is published as a printed book. We recommend this License principally for works whose purpose is instruction or reference.

1. APPLICABILITY AND DEFINITIONS

This License applies to any manual or other work, in any medium, that contains a notice placed by the copyright holder saying it can be distributed under the terms of this License. Such a notice grants a world-wide, royalty-free license, unlimited in duration, to use that work under the conditions stated herein. The "Document", below, refers to any such manual or work. Any member of the public is a licensee, and is addressed as "you". You accept the license if you copy, modify or distribute the work in a way requiring permission under copyright law.

A "Modified Version" of the Document means any work containing the Document or a portion of it, either copied verbatim, or with modifications and/or translated into another language.

A "Secondary Section" is a named appendix or a front-matter section of the Document that deals exclusively with the relationship of the publishers or authors of the Document to the Document's overall subject (or to related matters) and contains nothing that could fall directly within that overall subject. (Thus, if the Document is in part a textbook of mathematics, a Secondary Section may not explain any mathematics.) The relationship could be a matter of historical connection with the subject or with related matters, or of legal, commercial, philosophical, ethical or political position regarding them.

The "Invariant Sections" are certain Secondary Sections whose titles are designated, as being those of Invariant Sections, in the notice that says that the Document is released under this License. If a section does not fit the above definition of Secondary then it is not allowed to be designated as Invariant. The Document may contain zero Invariant Sections. If the Document does not identify any Invariant Sections then there are none.

The "Cover Texts" are certain short passages of text that are listed, as Front-Cover Texts or Back-Cover Texts, in the notice that says that the Document is released under this License. A Front-Cover Text may be at most 5 words, and a Back-Cover Text may be at most 25 words.

A "Transparent" copy of the Document means a machine-readable copy, represented in a format whose specification is available to the general public, that is suitable for revising the document straightforwardly with generic text editors or (for images composed of pixels) generic paint programs or (for drawings) some widely available drawing editor, and that is suitable for input to text formatters or for automatic translation to a variety of formats suitable for input to text formatters. A copy made in an otherwise Transparent file format whose markup, or absence of markup, has been arranged to thwart or discourage subsequent modification by readers is not Transparent. An image format is not Transparent if used for any substantial amount of text. A copy that is not "Transparent" is called "Opaque".

Examples of suitable formats for Transparent copies include plain ASCII without markup, Texinfo input format, LaTeX input format, SGML or XML using a publicly available DTD, and standard-conforming simple HTML, PostScript or PDF designed for human modification. Examples of transparent image formats include PNG, XCF and JPG. Opaque formats include proprietary formats that can be read and edited only by proprietary word processors, SGML or XML for which the DTD and/or processing tools are not generally available, and the machine-generated HTML, PostScript or PDF produced by some word processors for output purposes only.

The "Title Page" means, for a printed book, the title page itself, plus such following pages as are needed to hold, legibly, the material this License requires to appear in the title page. For works in formats which do not have any title page as such, "Title Page" means the text near the most prominent appearance of the work's title, preceding the beginning of the body of the text.

The "publisher" means any person or entity that distributes copies of the Document to the public.

A section "Entitled XYZ" means a named subunit of the Document whose title either is precisely XYZ or contains XYZ in parentheses following text that translates XYZ in another language. (Here XYZ stands for a specific section name mentioned below, such as "Acknowledgements", "Dedications", "Endorsements", or "History".) To "Preserve the Title" of such a section when you modify the Document means that it remains a section "Entitled XYZ" according to this definition.

The Document may include Warranty Disclaimers next to the notice which states that this License applies to the Document. These Warranty Disclaimers are considered to be included by reference in this License, but only as regards disclaiming warranties: any other implication that these Warranty Disclaimers may have is void and has no effect on the meaning of this License.

2. VERBATIM COPYING

You may copy and distribute the Document in any medium, either commercially or noncommercially, provided that this License, the copyright notices, and the license notice saying this License applies to the Document are reproduced in all copies, and that you add no other conditions whatsoever to those of this License. You may not use technical measures to obstruct or control the reading or further copying of the copies you make or distribute. However, you may accept compensation in exchange for copies. If you distribute a large enough number of copies you must also follow the conditions in section 3.

You may also lend copies, under the same conditions stated above, and you may publicly display copies.

3. COPYING IN QUANTITY

If you publish printed copies (or copies in media that commonly have printed covers) of the Document, numbering more than 100, and the Document's license notice requires Cover Texts, you must enclose the copies in covers that carry, clearly and legibly, all these Cover Texts: Front-Cover Texts on the front cover, and Back-Cover Texts on the back cover. Both covers must also clearly and legibly identify you as the publisher of these copies. The front cover must present the full title with all words of the title equally prominent and visible. You may add other material on the covers in addition. Copying with changes limited to the covers, as long as they preserve the title of the Document and satisfy these conditions, can be treated as verbatim copying in other respects.

If the required texts for either cover are too voluminous to fit legibly, you should put the first ones listed (as many as fit reasonably) on the actual cover, and continue the rest onto adjacent pages.

If you publish or distribute Opaque copies of the Document numbering more than 100, you must either include a machine-readable Transparent copy along with each Opaque copy, or state in or with each Opaque copy a computer-network location from which the general network-using public has access to download using public-standard network protocols a complete Transparent copy of the Document, free of added material. If you use the latter option, you must take reasonably prudent steps, when you begin distribution of Opaque copies in quantity, to ensure that this Transparent copy will remain thus accessible at the stated location until at least one year after the last time you distribute an Opaque copy (directly or through your agents or retailers) of that edition to the public.

It is requested, but not required, that you contact the authors of the Document well before redistributing any large number of copies, to give them a chance to provide you with an updated version of the Document.

4. MODIFICATIONS

You may copy and distribute a Modified Version of the Document under the conditions of sections 2 and 3 above, provided that you release the Modified Version under precisely this License, with the Modified Version filling the role of the Document, thus licensing distribution and modification of the Modified Version to whoever possesses a copy of it. In addition, you must do these things in the Modified Version:

A. Use in the Title Page (and on the covers, if any) a title distinct from that of the Document, and from those of previous versions (which should, if there were any, be listed in the History section of the Document). You may use the same title as a previous version if the original publisher of that version gives permission.

B. List on the Title Page, as authors, one or more persons or entities responsible for authorship of the modifications in the Modified Version, together with at least five of the principal authors of the Document (all of its principal authors, if it has fewer than five), unless they release you from this requirement.

C. State on the Title page the name of the publisher of the Modified Version, as the publisher.

D. Preserve all the copyright notices of the Document.

E. Add an appropriate copyright notice for your modifications adjacent to the other copyright notices.

F. Include, immediately after the copyright notices, a license notice giving the public permission to use the Modified Version under the terms of this License, in the form shown in the Addendum below.

G. Preserve in that license notice the full lists of Invariant Sections and required Cover Texts given in the Document's license notice.

H. Include an unaltered copy of this License.

I. Preserve the section Entitled "History", Preserve its Title, and add to it an item stating at least the title, year, new authors, and publisher of the Modified Version as given on the Title Page. If there is no section Entitled "History" in the Document, create one stating the title, year, authors, and publisher of the Document as given on its Title Page, then add an item describing the Modified Version as stated in the previous sentence.

J. Preserve the network location, if any, given in the Document for public access to a Transparent copy of the Document, and likewise the network locations given in the Document for previous versions it was based on. These may be placed in the "History" section. You may omit a network location for a work that was published at least four years before the Document itself, or if the original publisher of the version it refers to gives permission.

K. For any section Entitled "Acknowledgements" or "Dedications", Preserve the Title of the section, and preserve in the section all the substance and tone of each of the contributor acknowledgements and/or dedications given therein.

L. Preserve all the Invariant Sections of the Document, unaltered in their text and in their titles. Section numbers or the equivalent are not considered part of the section titles.

M. Delete any section Entitled "Endorsements". Such a section may not be included in the Modified Version.

N. Do not retitle any existing section to be Entitled "Endorsements" or to conflict in title with any Invariant Section.

O. Preserve any Warranty Disclaimers.

If the Modified Version includes new front-matter sections or appendices that qualify as Secondary Sections and contain no material copied from the Document, you may at your option designate some or all of these sections as invariant. To do this, add their titles to the list of Invariant Sections in the Modified Version's license notice. These titles must be distinct from any other section titles.

You may add a section Entitled "Endorsements", provided it contains nothing but endorsements of your Modified Version by various parties—for example, statements of peer review or that the text has been approved by an organization as the authoritative definition of a standard.

You may add a passage of up to five words as a Front-Cover Text, and a passage of up to 25 words as a Back-Cover Text, to the end of the list of Cover Texts in the Modified Version. Only one passage of Front-Cover Text and one of Back-Cover Text may be added by (or through arrangements made by) any one entity. If the Document already includes a cover text for the same cover, previously added by you or by arrangement made by the same entity you are acting on behalf of, you may not add another; but you may replace the old one, on explicit permission from the previous publisher that added the old one.

The author(s) and publisher(s) of the Document do not by this License give permission to use their names for publicity for or to assert or imply endorsement of any Modified Version.

5. COMBINING DOCUMENTS

You may combine the Document with other documents released under this License, under the terms defined in section 4 above for modified versions, provided that you include in the combination all of the Invariant Sections of all of the original documents, unmodified, and list them all as Invariant Sections of your combined work in its license notice, and that you preserve all their Warranty Disclaimers.

The combined work need only contain one copy of this License, and multiple identical Invariant Sections may be replaced with a single copy. If there are multiple Invariant Sections with the same name but different contents, make the title of each such section unique by adding at the end of it, in parentheses, the name of the original author or publisher of that section if known, or else a unique number. Make the same adjustment to the section titles in the list of Invariant Sections in the license notice of the combined work.

In the combination, you must combine any sections Entitled "History" in the various original documents, forming one section Entitled "History"; likewise combine any sections Entitled "Acknowledgements", and any sections Entitled "Dedications". You must delete all sections Entitled "Endorsements."

6. COLLECTIONS OF DOCUMENTS

You may make a collection consisting of the Document and other documents released under this License, and replace the individual copies of this License in the various documents with a single copy that is included in the collection, provided that you follow the rules of this License for verbatim copying of each of the documents in all other respects.

You may extract a single document from such a collection, and distribute it individually under this License, provided you insert a copy of this License into the extracted document, and follow this License in all other respects regarding verbatim copying of that document.

7. AGGREGATION WITH INDEPENDENT WORKS

A compilation of the Document or its derivatives with other separate and independent documents or works, in or on a volume of a storage or distribution medium, is called an "aggregate" if the copyright resulting from the compilation is not used to limit the legal rights of the compilation's users beyond what the individual works permit. When the Document is included in an aggregate, this License does not apply to the other works in the aggregate which are not themselves derivative works of the Document.

If the Cover Text requirement of section 3 is applicable to these copies of the Document, then if the Document is less than one half of the entire aggregate, the Document's Cover Texts may be placed on covers that bracket the Document within the aggregate, or the electronic equivalent of covers if the Document is in electronic form. Otherwise they must appear on printed covers that bracket the whole aggregate.

8. TRANSLATION

Translation is considered a kind of modification, so you may distribute translations of the Document under the terms of section 4. Replacing Invariant Sections with translations requires special permission from their copyright holders, but you may include translations of some or all Invariant Sections in addition to the original versions of these Invariant Sections. You may include a translation of this License, and all the license notices in the Document, and any Warranty Disclaimers, provided that you also include the original English version of this License and the original versions of those notices and disclaimers. In case of a disagreement between the translation and the original version of this License or a notice or disclaimer, the original version will prevail.

If a section in the Document is Entitled "Acknowledgements", "Dedications", or "History", the requirement (section 4) to Preserve its Title (section 1) will typically require changing the actual title.

9. TERMINATION

You may not copy, modify, sublicense, or distribute the Document except as expressly provided under this License. Any attempt otherwise to copy, modify, sublicense, or distribute it is void, and will automatically terminate your rights under this License.

However, if you cease all violation of this License, then your license from a particular copyright holder is reinstated (a) provisionally, unless and until the copyright holder explicitly and finally terminates your license, and (b) permanently, if the copyright holder fails to notify you of the violation by some reasonable means prior to 60 days after the cessation.

Moreover, your license from a particular copyright holder is reinstated permanently if the copyright holder notifies you of the violation by some reasonable means, this is the first time you have received notice of violation of this License (for any work) from that copyright holder, and you cure the violation prior to 30 days after your receipt of the notice.

Termination of your rights under this section does not terminate the licenses of parties who have received copies or rights from you under this License. If your rights have been terminated and not permanently reinstated, receipt of a copy of some or all of the same material does not give you any rights to use it.

10. FUTURE REVISIONS OF THIS LICENSE

The Free Software Foundation may publish new, revised versions of the GNU Free Documentation License from time to time. Such new versions will be similar in spirit to the present version, but may differ in detail to address new problems or concerns. See http://www.gnu.org/copyleft/.

Each version of the License is given a distinguishing version number. If the Document specifies that a particular numbered version of this License "or any later version" applies to it, you have the option of following the terms and conditions either of that specified version or of any later version that has been published (not as a draft) by the Free Software Foundation. If the Document does not specify a version number of this License, you may choose any version ever published (not as a draft) by the Free Software Foundation. If the Document specifies that a proxy can decide which future versions of this License can be used, that proxy's public statement of acceptance of a version permanently authorizes you to choose that version for the Document.

11. RELICENSING

"Massive Multiauthor Collaboration Site" (or "MMC Site") means any World Wide Web server that publishes copyrightable works and also provides prominent facilities for anybody to edit those works. A public wiki that anybody can edit is an example of such a server. A "Massive Multiauthor Collaboration" (or "MMC") contained in the site means any set of copyrightable works thus published on the MMC site.

"CC-BY-SA" means the Creative Commons Attribution-Share Alike 3.0 license published by Creative Commons Corporation, a not-for-profit corporation with a principal place of business in San Francisco, California, as well as future copyleft versions of that license published by that same organization.

"Incorporate" means to publish or republish a Document, in whole or in part, as part of another Document.

An MMC is "eligible for relicensing" if it is licensed under this License, and if all works that were first published under this License somewhere other than this MMC, and subsequently incorporated in whole or in part into the MMC, (1) had no cover texts or invariant sections, and (2) were thus incorporated prior to November 1, 2008.

The operator of an MMC Site may republish an MMC contained in the site under CC-BY-SA on the same site at any time before August 1, 2009, provided the MMC is eligible for relicensing.

ADDENDUM: How to use this License for your documents

To use this License in a document you have written, include a copy of the License in the document and put the following copyright and license notices just after the title page:

```
Copyright (C)  year  your name.
Permission is granted to copy, distribute and/or modify this document
under the terms of the GNU Free Documentation License, Version 1.3
or any later version published by the Free Software Foundation;
with no Invariant Sections, no Front-Cover Texts, and no Back-Cover
Texts.  A copy of the license is included in the section entitled ``GNU
Free Documentation License''.
```

If you have Invariant Sections, Front-Cover Texts and Back-Cover Texts, replace the "with...Texts." line with this:

```
with the Invariant Sections being list their titles, with
the Front-Cover Texts being list, and with the Back-Cover Texts
being list.
```

If you have Invariant Sections without Cover Texts, or some other combination of the three, merge those two alternatives to suit the situation.

If your document contains nontrivial examples of program code, we recommend releasing these examples in parallel under your choice of free software license, such as the GNU General Public License, to permit their use in free software.